OXFORD STUDIES

IN

NUCLEAR PHYSICS

GENERAL EDITOR

P. E. HODGSON

NUCLEAR
STATISTICAL
SPECTROSCOPY

S. S. M. Wong

University of Toronto

OXFORD UNIVERSITY PRESS • NEW YORK

CLARENDON PRESS • OXFORD

1986

Oxford University Press

Oxford   New York   Toronto
Delhi Bombay Calcutta Madras Karachi
Petaling Jaya Singapore Hong Kong Tokyo
Nairobi Dar es Salaam Cape Town
Melbourne Auckland

and associated companies in
Beirut Berlin Ibada Nicosia

Copyright © 1986 by Samuel Shaw Ming Wong

Published by Oxford University Press, Inc., 200 Madison Avenue,
New York, New York 10016

Oxford is the registered trademark of Oxford University.

Library of Congress Cataloging-in-Publication Data

Wong, S. S. M. (Samuel Shaw Ming), 1937–
    Nuclear statistical spectroscopy.

    (Oxford studies in nuclear physics)
    Bibliography: p.
    Includes index.
        1. Nuclear spectroscopy. 2. Nuclear structure --
    Statistical methods. 3. Nuclear spin--Statistical
    methods. 4. Eigenvalues. 5. Hamiltonian systems.
    I. Title. II. Title: Statistical spectroscopy.
    III. Series.
    QC454.N8W66 1986   539.7   85-18917
    ISBN 0-19-504004-X

British Library Cataloguing in Publication Data

Wong, S.S.M.
    Nuclear statistical spectroscopy. — (Oxford
    studies in nuclear physics)
    1. Nuclear spectroscopy
    I. Title
    539.7'44   QC454.N8
    ISBN 0-19-504004-X

Printing (last digit): 9 8 7 6 5 4 3 2 1
Printed in the United States of America

Preface

This monograph was started as an extension of a part of the lecture notes given at the Winter School in Nuclear Physics held in Peking, 1980 − 1. It seems to be a good time for writing such a volume since the subject of statistical spectroscopy has developed to a stage where a concise account may be useful.

In the fifties and early sixties, it was generally believed that the advances in computing would enable one to carry out shell model and other similar calculations on a scale large enough to provide us with a good microscopic understanding of the nucleus. Indeed, large scale calculations have taught us a lot about the nucleus but we are still very far from reaching the elusive goal of being able to carry out a large "enough" calculation. In fact, one of the important lessons learned from these calculations is that an alternative to the eigenvalue-problem approach underlying most of the microscopic studies should be attempted.

Statistical spectroscopy was developed as a marriage between the shell-model type of *microscopic* approaches and the purely statistical approaches used in certain nuclear reaction theories and random matrix studies. Statistical techniques are employed to reduce the complexity of calculations without sacrificing the essential information in the results. This is a tall order, and can be achieved only by departing somewhat from the traditional ways of carrying out nuclear structure investigations. Statistical spectroscopy differs from purely statistical approaches by keeping a direct contact with the nuclear Hamiltonian at all times. It is also different from traditional microscopic approaches since individual eigenstates are sacrificed in favour of the dependence of a physical quantity on energy and other variables.

Being a mixture of two different traditional approaches, statistical spectroscopy is often misunderstood, especially when some of the results produced are unfamiliar in either approach. Furthermore, a new set of techniques was developed and some new ways of examining problems were brought into the study of nuclei. Partly as a result of these innovations, the subject is often described in a language that is sometimes strange to most nuclear physicists.

There are therefore three basic aims of this monograph. The first is to introduce the subject of statistical spectroscopy as a new tool for nuclear structure studies and in a way that is easy for the uninitiated reader. The second is to bring most of the important techniques into one place so that it can serve as a reference for a person wishing to start using the methods. The third is to show that the new outlook provided by statistical spectroscopy, although not yet fully developed, has potentials that are worthwhile examining. To avoid deviating from these aims, there is no attempt made to give a complete review of the subject. Several topics are omitted so as to keep the account short and to avoid diverting the reader from the central theme. Even the list of references is reduced to a minimum by ignoring earlier developments in favour of later accounts that are more complete.

The author is greatly indebted to his close collaborators, in particular, Bei-dwo Chang, Jerry Draayer, and Bruce French. Special thanks must go to Jan van Kranendonk for tirelessly reading over the manuscripts numerous times and making many valuable suggestions.

This monograph is produced from a camera-ready copy typeset on the University of Toronto Physics/Astronomy computing facilities.

Toronto   S. S. M. Wong
July 1985

Contents

NUCLEAR
STATISTICAL
SPECTROSCOPY

Chapter I

Introduction

A finite nucleus consists of a fixed number of nucleons each of which moves in the average one-body field generated by all the other nucleons. In addition, the nucleons also interact with each other through a residual two-body interaction. In such a many-body system, the wave functions for the system of nucleons are usually constructed as linear combinations of antisymmetrized products of single-particle wave functions. The Hilbert space for these many-particle states is in principle infinite: however, for practical reasons, calculations are carried out in finite spaces defined by a set of single-particle states. Since it is usually the low-energy part of the spectrum of a nuclear system that is of interest, only a limited number of the single-particle states near the Fermi energy are considered to be active. The states below these active ones are filled by particles which are not allowed to be excited since such excitations would require large amounts of energy. Similarly, excitation of particles into the states above the active ones are forbidden. Except for their influence on the *effective* Hamiltonian in the active space, all the single-particle states other than the active ones can therefore essentially be ignored.

Within this finite many-particle space, calculations of a physical quantity are restricted to its contributions in the space used. For example, the density of states is generally an increasing function of the energy simply from the fact that more single-particle states are accessible to the system at higher energies. On the other hand, a calculation of the density of states using a finite space will produce a function that must eventually decrease with increasing energy and go to zero asymptotically since the total number of states in the finite space is limited. This unrealistic feature of calculations using a finite space causes no problem when we compare the results with experiment if we assume that the space used is sufficiently large to encompass the region of interest. It is therefore understood that the results we calculate are always the *partial* results in a finite space. Following common practice, we shall refer to this part of the Hilbert space as the *active space*, or the "space" for short.

In statistical spectroscopy we deal with the generalized function, or the distribution, that describes the dependence of a physical quantity on energy and other variables. This is different from the usual approach in which the calculated results are the expectation values of the corresponding operator over specific states or the transition strengths between particular pairs of states obtained by solving an eigenvalue problem. The advantage of using distributions is that, since the partial result of a quantity in a finite space is bounded, the energy and other dependences can be expanded in terms of moments. If the expansions are restricted to low orders, the moments involved are then traces of simple products of operators and they are in general far easier to obtain than to solve the eigenvalue problem in a large space.

The common aim of many studies in nuclear physics is to understand the nucleus starting from the fundamental nucleon-nucleon interaction. One of the problems encountered is that very large spaces must be used before the results can be compared with experiments. On the other hand, most of the work involved may be superfluous since only a small part of the information generated by such calculations is actually used. For example, when the Hamiltonian matrix is diagonalized in a space of several thousand basis states, often only the lowest few eigenstates are of interest. Furthermore the eigenvectors, each consisting of thousands of components, are used in general to obtain only a few expectation values and transition matrix elements. Instead of discarding at the end most of the details one cannot make use of, it would be far more profitable not to calculate them. From a practical point of view, such a procedure is essential. The size of the many-particle space grows exponentially with the addition of single-particle states to the active space. The need to increase the number of single-particle states is quite obvious from experimental evidence, and no improvement in computational techniques can hope to cope with the problem of exponential growth in the dimension of the many-particle space unless a new approach is taken. Statistical spectroscopy represents one such attempt.

In a nucleus, the energy dependence of the expectation values or the excitation strengths of an operator can be separated into two parts: a secular part corresponding to the slow changes that are noticeable only over a distance of many states, and fluctuations corresponding to differences between neighbouring states. In statistical spectroscopy, the same separation can also be characterized in terms of the moments of the corresponding distributions, the low-order moments describing the secular variation and the high-order ones the fluctuations. The expected economy comes from the belief that the slow variations of a distribution are the important features of the system and that, consequently, an expansion of the distribution can be limited to the low-order moments.

The justification for adopting such a point of view comes in part from studies using ensembles of random matrices. It has been shown that fluctuations in the distribution of a physical quantity are the properties common to many systems and are therefore not useful for understanding specific systems such as nuclei. As a result, a large part of the complexity in microscopic calculations in large spaces can be avoided without any loss of essential information.

The fundamental reason that statistical spectroscopy is meaningful in the study of the nuclear system comes from the large number of independent degrees of freedom present. In such a system, the effect of the central limit theorem dominates; the distributions of most observables are essentially Gaussian and are determined by a few low-order moments. The role played by the higher-order moments, conveyed by the details generated in large microscopic calculations, is reduced when the system is dominated by statistical properties. Consequently, it is mostly the low-order moments of the distributions that can tell us something about the nucleus.

Certainly the statistical point of view cannot be taken to the extreme. There are many features of the nuclear system that are clearly not statistical in nature. For example, aspects of a nucleus involving the coherent motion of many nucleons cannot be treated with advantage using statistical spectroscopy. Furthermore, if the interest is in a particular state, because of certain properties that distinguish that state from its neighbours, a statistical treatment is not suitable. Many models have been designed to understand such features successfully; it would not be appropriate for statistical spectroscopy to compete in such areas.

This monograph is divided into three parts. The next two chapters are concerned with the general background of the statistical approach and its conceptual dependence on random matrices. Chapter II gives a review of those aspects of random matrix studies that are relevant to the subject of statistical spectroscopy. The main point is to establish that fluctuations in energy levels and transition strengths are the properties of quantum mechanical systems in general from which we cannot expect to learn anything about the nuclear system. This is done by applying statistical measures to fluctuations and observing that there is no difference between the results obtained from random matrices, shell model calculations, and experimental values. The results lead to the conclusion that the higher-order moments of a distribution are not useful for understanding the nuclear system and that it is more profitable to concentrate on the low-order moments. Chapter III gives the background in statistics required for statistical spectroscopy. The distributions of interest in nuclear structure studies can be classified into three categories corresponding to energy levels, expectation values, and excitation strengths. In each case, the low-order moments are the traces of products of the Hamiltonian and excitation operators. Expressions for these product operators are given in Chapter III.

The various methods of calculating the traces of these product operators are given in Chapters IV to IX. Chapter IV uses elementary considerations involving combinatorial arguments to derive the relations between moments of simple operators, such as the powers of the Hamiltonian, and the matrix elements that define the operator. For traces of more complicated operators, a diagrammatic method is developed in Chapter V. Instead of taking the trace over all the states in the many-body space, it is necessary for many purposes to subdivide the active space into subspaces. In Chapter VI, the diagrammatic method is extended to traces over configuration subspaces defined by definite numbers of particles in single-particle orbits. Different methods are required to obtain the traces in other types of subspace, and they are given in the next three chapters. The projection of moments for subspaces with definite isospin is covered in Chapter VII. For subspaces with fixed spin, there is no easy way available yet to calculate the moments. Three different methods are given in Chapter VIII to illustrate the possibilities. As for the irreducible representations of an arbitrary group, no general method is known that can produce the moments. A brief discussion of the difficulties involved in this problem and of the available techniques is given in Chapter IX.

The applications to various physical problems given in Chapter X are selected with a view to demonstrate the new outlook statistical spectroscopy can provide. Although standard nuclear spectroscopy quantities such as energy level positions, static moments, and transition rates can be obtained, it would be incorrect to require statistical spectroscopy to produce results for these quantities with comparable accuracy as models designed for these studies. The strength of statistical spectroscopy is in spaces far larger than can be handled by conventional microscopic models, such as the nuclear shell model, and in providing new ways to examine nuclear structure problems. It is hoped that the examples given can provide some guidance for the possibilities.

The six appendices give some of the specific technical details that are not necessarily central to statistical spectroscopy but are likely to be useful in applying the methods.

Chapter II

Random matrix studies

The statistical approach to nuclear structure is based on two premises. The first one is that a separation can be made of the roles played by low- and high-order moments, the low-order ones being responsible for the slow variations in the distribution as a function of energy and the high-order ones for the fluctuations. The second one is that the information of interest lies mainly in the smooth variations. Neither of these premises can be established firmly, but random matrix studies provide strong support for their validity under a reasonable set of assumptions.

II.1 Ensemble of Hamiltonians

In a given space, the properties of a system are governed by the effective Hamiltonian operating in the space. However, since the interaction between nucleons inside a nucleus is not completely known, one must refrain from drawing general conclusions based on a particular Hamiltonian. Furthermore, since we are interested here in the general features of nuclei, we are not concerned with the special characteristics of a few nuclei or properties resulting from the peculiarities of the effective Hamiltonian operating in the region. This calls for the introduction of ensembles in analogy with the ensembles used in statistical mechanics. If Q is the physical quantity of interest, it will be calculated with the eigenvectors obtained from solving the Schrödinger equation with all "reasonable" Hamiltonians. By "reasonable" Hamiltonian we mean here one that satisfies all the well known properties of a nuclear Hamiltonian, such as time reversal invariance, rotational symmetry, and consisting of one- and two-body interactions. The term includes, but is not restricted to, *realistic* Hamiltonians, which usually means either Hamiltonians derived from free nucleon-nucleon scattering data, or Hamiltonians whose defining matrix elements are obtained by fitting to experimental information of nuclei.

Let $\hat{Q}$ represent the operator corresponding to the physical quantity Q of interest to us. In general, the values of Q obtained, say, in the form of the expectation values of $\hat{Q}$, are different for different Hamiltonians. If the results calculated with the eigenvectors of all "reasonable" Hamiltonians are clustered in a narrow region, we can safely assume that the average over the collection, or *ensemble*, of results provides a good estimate of the value of Q calculated using eigenvectors obtained with the true Hamiltonian.

The approach is, however, different from conventional statistical mechanics which works with the time-development of a system under the action of a fixed Hamiltonian. Here we have a fixed system but different Hamiltonians. Instead of assuming that the system is ergodic in time, *i.e.*, given sufficient time the system will, with equal probability, be in all possible states each of which is represented by a member of the ensemble, we assume that each "reasonable" Hamiltonian used to calculate the ensemble result of Q is equally representative of the true Hamiltonian. The proof for this type of ergodicity is not any easier than in statistical mechanics (see Brody *et al.*, 1981). On the other hand, if the ensemble distribution is narrow, it is highly probable that the ensemble-averaged value is representative of what one would obtain using the true Hamiltonian, since all the "reasonable" Hamiltonians give similar results in this case. On the other hand, if the ensemble distribution is flat, the ensemble average does not provide us with any guide concerning the possible outcome with the true Hamiltonian. There may be several reasons for this failure and one of them may well be that we had chosen the wrong ensemble.

It is therefore prudent to examine the following two points before drawing any conclusions based on the ensemble results. In the first place, we must ensure that the ensemble distribution is narrow so that it is unlikely to find values far away from the average. This can be done by evaluating, in addition to the mean, also the variance of the ensemble distribution, the ensemble average of the square of the quantity minus the square of the ensemble mean. A small variance indicates that the ensemble distribution is narrow: however, as usual, higher moments are required to specify the shape of the distribution further but these are generally much harder to obtain in the case of ensemble distributions.

Since the true nuclear Hamiltonian is not known, it is difficult to ensure that it is a member of the assumed ensemble. As a result, it may be tempting to enlarge the ensemble as much as possible by relaxing the conditions for a Hamiltonian to be a "reasonable" one. Furthermore, the ensemble must also be mathematically manageable. In general, the requirements of mathematical convenience and of "reasonable" Hamiltonians for the nuclear system do not necessarily coincide, and the temptation here is again to enlarge the ensemble so as to accommodate both requirements. On the other hand, if the ensemble is too large, the proportion of truly "reasonable" Hamiltonians may become so small that the ensemble averaged results will no longer be representative of the nuclear system. As we shall see in an example later, if the ensemble is dominated by "unreasonable" members, the average may not be physically meaningful even if the ensemble distribution is narrow.

II.2   Moments of a distribution

To approach the study of distributions and averages in a more quantitative manner, we shall first define the moments that characterize a distribution in general. Let $\rho(x)$ represent the distribution of a physical quantity as a function of the variable x. Unless stated otherwise, we shall normalize $\rho(x)$ to unity,

$$\int_{-\infty}^{+\infty} \rho(x)\, dx = 1. \tag{1}$$

For example, if we are discussing the distribution of energy levels as a function of the energy, $\rho(x)$ is the number of energy levels per unit interval at energy x divided by the total number of levels in the space.

In statistics $\rho(x)$ is referred to as the frequency function and the term distribution function is reserved for $F(y)$, the integrated strength of $\rho(x)$ up to $x = y$,

$$F(y) = \int_{-\infty}^{y} \rho(x)\, dx.$$

We shall, however, follow the practice more common in physics and call $\rho(x)$ the distribution and $F(y)$ the integrated strength.

A distribution $\rho(x)$ is characterized by its moments

$$M_\mu = \int_{-\infty}^{+\infty} x^\mu \rho(x)\, dx. \tag{2}$$

For discrete distributions, such as eigenvalue distributions

$$\rho(x) = \frac{1}{d} \sum_{i=1}^{d} \delta(x - E_i), \tag{3}$$

the integral in eq. (2) reduces to a sum,

$$M_\mu = \frac{1}{d} \sum_{i=1}^{d} (E_i)^\mu. \tag{4}$$

The first moment M_1 is also called the centroid or the mean of the distribution,

$$C = M_1.$$

It gives the location of the distribution on the x-axis. The second moment is related to the variance $\mathcal{V}$,

$$\mathcal{V} = M_2 - C^2 \equiv \sigma^2.$$

The square root, σ, of the variance characterizes the width of the distribution and is generally referred to as the standard deviation or *width*.

The values of the moments as defined in eq. (2) depend on the location of the origin and on the scale selected for the x-axis: the shape of the distribution is, on the other hand, invariant under a shift of the origin along the x-axis and under a change of its scale. It is therefore more convenient to use the normalized central moments,

$$M_\mu = \int_{-\infty}^{+\infty} \left(\frac{x-C}{\sigma}\right)^\mu \rho(x)\, dx \longrightarrow \frac{1}{d} \sum_{i=1}^{d} \left(\frac{E_i - C}{\sigma}\right)^\mu, \tag{5}$$

to characterize a distribution. Since the information concerning the location and scale is already contained in the centroid C and width σ, we have

$$M_1 = 0 \quad \text{and} \quad M_2 = 1.$$

The two sets of moments are related by

$$M_\mu = \frac{1}{\sigma^\nu} \sum_{\nu=0}^{\mu} \binom{\mu}{\nu} M_{\mu-\nu}(-C)^\nu, \tag{6}$$

as can be seen from eq. (5).

Instead of moments, cumulants κ_μ may also be used to characterize a distribution (Cramer, 1946). They are related to the moments by matching the coefficients associated with the same powers of t on the two sides of the equation

$$1 + \sum_{\mu=1}^{\infty} \frac{M_\mu}{\mu!}(it)^\mu = \exp\left\{\sum_{\mu=1}^{\infty} \frac{\kappa_\mu}{\mu!}(it)^\mu\right\}. \tag{7}$$

For the few lowest orders the explicit relations in terms of normalized central moments are

$$\begin{aligned}
\kappa_1 &= M_1 = 0, & \kappa_2 &= M_2 = 1, \\
\kappa_3 &= M_3, & \kappa_4 &= M_4 - 3M_2^2, \\
\kappa_5 &= M_5 - 10M_3, & \kappa_6 &= M_6 - 15M_4 - 10M_3^2 + 30.
\end{aligned} \tag{8}$$

The third cumulant, κ_3, is also known as the skewness since it measures the extent a distribution deviates from a symmetric one. The fourth cumulant, κ_4, is referred to as the excess. If it is positive, the distribution is flat compared with a Gaussian with the same centroid and width; if it is negative, it is sharper.

The Gaussian, or normal, distribution

$$\rho_G(x) = \frac{1}{\sqrt{2\pi}\sigma} \exp{-\frac{(x-C)^2}{2\sigma^2}}, \tag{9}$$

is one of the standard distributions that we shall be using often. It is completely defined by two moments, the centroid C and width σ. Since it is symmetric around $x = 0$, all the odd moments are zero and the even ones are given in terms of the width σ by

$$M_{2\mu} = \sigma^{2\mu}(2\mu - 1)!! = \sigma^{2\mu}(2\mu - 1)(2\mu - 3)(2\mu - 5)\cdots 3 \cdot 1. \qquad (10)$$

The cumulants, on the other hand, all vanish except κ_2 which we can take as unity for most purposes.

II.3   Random matrix ensemble

In spectroscopy, we have in general a space spanned by all the states formed by putting m particles in N single-particle states. This will be called the m-particle space. An operator pertaining to this space is completely specified if all its matrix elements in the space are given. For simplicity we shall consider here only operators with a definite particle rank k, *i.e.*, the operator vanishes in spaces with $m < k$ and all the matrix elements for $m > k$ can be obtained, or *propagated*, from those in the k-particle space alone. Such operators will be called k-body operators. More complicated operators with mixed particle ranks can also be studied, but we shall not do so in this chapter since this would not add anything of interest in random matrix studies.

A k-body operator is therefore defined by its matrix elements in the k-particle space. Other equivalent sets of defining matrix elements can be constructed; we shall however restrict ourselves in this chapter to the most straightforward set since the conclusions we wish to draw are independent of such choices. If the form of the operator is known, we can calculate all the defining matrix elements in the k-particle space. However this is not the case of interest to us here: we are concerned with operators that are not necessarily known to us in all detail and we wish to make some progress using a statistical approach.

As far as possible we should make use of what is known about the operator. For example, if

$$\hat{Q}(\varsigma) = a\,f(\varsigma) + b\,g(\varsigma) + c\,h(\varsigma),$$

where $f(\varsigma)$, $g(\varsigma)$, and $h(\varsigma)$ are known functions of ς but the values of a, b, and c are completely uncertain within some finite ranges, we can take any set of three possible values (a_i, b_i, c_i) within the ranges for the coefficients in the operator $\hat{Q}$. Each set then constitutes a possible member of the ensemble. To ensure as broad and unbiased a selection of possible members as possible, we can use a set of three random numbers within the allowed ranges for (a_i, b_i, c_i). Such a collection of different possible operators $\hat{Q}$, each corresponding to a different set of values (a_i, b_i, c_i), clearly satisfies all the conditions to form a statistical ensemble for our purposes described earlier.

On the other hand, if the form of the operator is not known, except perhaps that it is a k-body operator, we have no way to calculate the defining matrix elements for the operator. In other words, any set of k-body matrix elements forms a possible realization of the operator. To avoid any bias, we can use random numbers for the defining matrix elements. Each set of random numbers then forms a member of the ensemble for this operator in the k-particle space. On propagating the k-body matrix elements to the m-particle space for $m > k$, we obtain a member of the random matrix ensemble for this operator in the m-particle space.

If we have further knowledge about the operator, it can be incorporated as a constraint in the selection of random numbers for the defining matrix elements. For example, the nuclear interaction can be chosen to be hermitian, hence the Hamiltonian matrix is real and symmetric. This reduces the number of independent matrix elements required to define a matrix of dimension d from d^2 to $d(d+1)/2$. Further constraints, such as spin and isospin conservations may also be imposed and they can reduce the number of independent defining matrix elements further.

II.4   Gaussian orthogonal ensemble

Our interest is in an m-particle space for $m > k$. The Hamiltonian matrix elements in the m-particle space can be expressed in terms of those in the defining k-particle space. In general, the propagation from k- to m-particle space is a tedious process. More importantly, it also implies a set of constraints on the m-particle Hamiltonian matrix elements that are very difficult to handle analytically. Purely on the basis of mathematical convenience, it was suggested by Wigner in 1955 to replace the m-particle matrix elements directly by random numbers. For an m-particle space of dimension d, there are then $d(d+1)/2$ independent elements for a real symmetric matrix. In the original Hamiltonian matrix, the value of a particular element depends on the single-particle basis used: however, the distribution of all the elements in the matrix must be invariant under a rotation of the single-particle states. To maintain this invariance, the distribution of the random numbers chosen to represent these matrix elements must be Gaussian centered around zero. Since the matrix is symmetric, the width of the distribution of the diagonal matrix elements must be $\sqrt{2}$ times that for the off-diagonal matrix elements. The exact value of the width is immaterial here since it can be treated as the unit of energy. Such an ensemble is known as a Gaussian Orthogonal Ensemble or GOE for short.

It is well-known that the eigenvalue distribution $\rho(E)$, or density of states, for such an ensemble, shown in Fig. II-1, is semicircular in shape

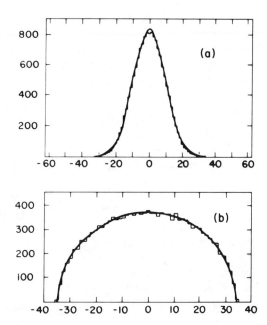

Fig. II-1. Histograms of level density distributions for two ensembles with 50 members in each. The TBRE (Two-Body Random Ensemble) results, shown in (a), are compared with a Gaussian curve, while the GOE (Gaussian Orthogonal Ensemble) results, shown in (b), are compared with a semicircular distribution. In both cases, the dimension of the matrix is 294.

(Wigner, 1967),

$$\rho(x) = \frac{1}{2\pi}\sqrt{4 - x^2}, \tag{11}$$

where $x = (E - C)/\sigma$, the energy measured from the origin in units of the width σ of the distribution. The odd moments of the distribution vanish and the even moments are given by

$$M_{2\mu} = \frac{1}{2\mu + 1}\binom{2\mu + 1}{\mu} = \frac{1}{\mu + 1}\binom{2\mu}{\mu}. \tag{12}$$

These numbers are also known as the Catalan numbers.

II.5   Two-body random ensemble

A nuclear Hamiltonian is usually a sum of one- and two-body operators. The one-body part comes from the kinetic energy of individual nucleon motion and binding energy to the average central potential provided by the rest of the nucleons, while the two-body part derives from the residual interaction. For model Hamiltonians in finite spaces, renormalization effects due to the truncation of the Hilbert space can in principle introduce terms of higher particle rank. Little is known about these higher-order terms but all indications seem to imply that their effects are unimportant. We shall therefore ignore them from now on.

The one-body part can be easily shown to produce a Gaussian state density distribution. Consider first the case with only two active orbits, $r = 1, 2$, each with the number of single-particle states $N_r \gg m$. The energy of a state E_{m_1, m_2} is simply given by the occupancies m_r and the single-particle energies ϵ_r of the orbits,

$$E_{m_1, m_2} = \sum_{r=1}^{2} m_r \epsilon_r, \tag{13}$$

with $m_1 + m_2 = m$. The number of states in the configuration (m_1, m_2) is

$$d_{m_1 m_2} = \binom{N_1}{m_1} \binom{N_2}{m_2}, \tag{14}$$

and they all have the same energy $E_{m_1 m_2}$. The number of states in the energy interval ΔE for this simple case is given by the sum of the dimensions of all the configurations in ΔE, and the state density is obtained from it upon dividing by ΔE. For $N_1 = N_2$, the case is identical to a binomial distribution which, as $N_1, N_2 \to \infty$, goes to a Gaussian distribution. For more than two orbits, a multinomial distribution takes the place of the binomial distribution and, unless the single-particle energy distribution is a highly non-uniform one, the distribution goes also to a Gaussian distribution in the limit of large N_r. For more detailed discussions, see e.g., Gervois (1972).

The spectrum of a $k = 1$ operator is of little interest since all the states in a given configuration are degenerate. This does not happen in real nuclei because of the action of the two-body interaction. Except for the spin-isospin invariance of the nuclear Hamiltonian, which groups all $(2J + 1)(2T + 1)$ states for given (J, T) together, nuclear spectra display both slow variations with energy as well as rapid fluctuations. Hence it is essential to include the $k = 2$ part of the Hamiltonian in any realistic study of the nucleus, especially if we wish to learn something concerning the fluctuation properties.

When spin and isospin are taken as good quantum numbers, the defining space of a two-body Hamiltonian has a blocked structure with zeros for all the off-diagonal matrix elements between states with different J or T. For real and symmetric Hamiltonians, we have the further condition that $H_{ij} = H_{ji}$. The number of independent defining matrix elements in the space is therefore given by $\sum_{JT} d_{JT}(d_{JT} + 1)/2$, where d_{JT} is the number of k-particle states with spin-isospin (JT) and any allowed values of (MZ), the projections of (JT) on the quantization axes. A random ensemble can be constructed by replacing the independent defining two-body matrix elements by random numbers. The one-body part is usually ignored, since for $m \gg 2$ the relative importance of the contribution from the one-body part is much less than that of the two-body part. Such an ensemble is generally known as the Two-Body Random Ensemble or TBRE for short.

TBRE must be realized in a space with an explicit single-particle basis. However, the conclusions are independent of the particular basis chosen since we have no reason to prefer one set of single-particle orbits over another one. From a practical point of view, this means that the number of active single-particle orbits must not be small.

The m-particle states must be antisymmetrized in order to satisfy the Pauli exclusion principle. This requirement makes it difficult, if not impossible, to propagate analytically the Hamiltonian matrix elements for finite N from the defining k-particle space to the required m-particle space. As a result, most of the results for TBRE are obtained numerically using shell model techniques. In principle this is not a problem, except that the cost of calculating a large number of matrices of high dimension becomes rather prohibitive. Furthermore, one must isolate effects due to the finite size of the space used, otherwise the conclusions may not be valid in a different space.

II.6   Gaussian versus semicircular eigenvalue distribution

It is well known that the eigenvalue distribution of TBRE is Gaussian while that for GOE is semicircular. Fig. II-1 shows a numerical simulation of the situation with 294-dimensional matrices. The difference is now well understood and, as we shall see later, it does not seem to affect the fluctuation properties.

We shall consider the more general case of a k-body Hamiltonian; the $k = 2$ case then corresponds to that obtained numerically in the previous section. Ignoring angular momentum and other selection rule considerations, the number of defining matrix elements for a hermitian, k-body operator is $d_k(d_k + 1)/2$, where d_k is the dimension of the k-particle space. Our aim is to evaluate in the m-particle space the moments of the Hamiltonian, and from the values of these moments deduce the distribution of the eigenvalues. Using the angle brackets $\langle \ \rangle$ to represent the average trace, *i.e.*, the trace divided by the dimension of the space, we can write the moment of order μ of the Hamiltonian in the form

$$\langle HHHH \cdots\cdots HH \rangle.$$
$$|\leftarrow \mu \text{ in number} \rightarrow |$$

Each factor H can be expressed in terms of the defining matrix elements $W_{\alpha\beta}$,

$$H = \sum_{\alpha\beta} W_{\alpha\beta} Z_\alpha^\dagger Z_\beta, \tag{15}$$

where $Z_\alpha^\dagger$ and Z_β are, respectively, the creation operator for the k-particle state α and the annihilation operator for the k-particle state β. For random matrix studies $W_{\alpha\beta}$ are random numbers which, without loss of generality, can be chosen to have a Gaussian distribution with centroid at zero. This ensures that the distributions are centered at the origin when averaged over the entire ensemble. The particular form of the distribution of the random numbers, however, has very little bearing on the following arguments.

The effect of a k-body operator acting on an m-particle state for $m \geq k$ is to change the single-particle states for k of them. In order for the trace to be nonvanishing, the action of all μ Hamiltonians must return the m-particle state to the one we started with. In second quantized language, this is equivalent to the requirement that all the operators are fully contracted among themselves. There are many ways to achieve this, including some with intermediate states identical with the starting state. This means that some of the μ Hamiltonians form separate fully-contracted clusters. Using the same letters to indicate those H's belonging to the same cluster, we can write the various contributing terms to the moment of order μ as $\langle AABBCBCCC \cdots \rangle$, $\langle ABABBC \cdots \rangle$, etc. For example, $\langle HHH \rangle$ has the possibilities $\langle AAA \rangle$, $\langle AAB \rangle$, $\langle ABA \rangle$, $\langle BAA \rangle$, and $\langle ABC \rangle$. Since each cluster is fully contracted within itself, the average trace of the entire product of Hamiltonians can be decomposed into a product of traces each of which involves only a single cluster,

$$\langle AABBBCC \cdots \rangle \longrightarrow \langle AA \cdots \rangle \langle BBB \cdots \rangle \langle CC \cdots \rangle \langle \cdots \rangle.$$

Furthermore, since H is centered, $\langle H \rangle = 0$. Thus terms involving clusters of only one H vanish.

Next we shall argue that asymptotically, as the number of single-particle states $N \to \infty$, contributions from terms involving only products of clusters of two H's will dominate the trace. First of all,

$$\langle AA \rangle = \sum_{\alpha\beta} W_{\alpha\beta} W_{\beta\alpha} = \sum_{\alpha\beta} W_{\alpha\beta}^2, \tag{16}$$

since $W_{\alpha\beta} = W_{\beta\alpha}$ by the hermitian property of the Hamiltonian. Being a sum of squares, the trace of a cluster of two Hamiltonians has a non-vanishing positive value. On the other hand, the higher-order clusters are predominately made up of terms involving products of uncorrelated random numbers; they must vanish on ensemble averaging. The only contributions come from relatively small numbers of products involving only squares of random numbers. For example in $\langle AAAA \rangle$ there are many more products of the form $W_{\alpha\beta} W_{\beta\gamma} W_{\gamma\delta} W_{\delta\alpha}$, which will vanish on ensemble averaging, than products of the form $W_{\alpha\beta} W_{\beta\gamma} W_{\gamma\beta} W_{\beta\alpha}$, which are positive definite and will therefore contribute to the ensemble average. Terms made up of clusters of two are, however, always positive in value and, as a result, their importance will dominate the trace as $N \to \infty$. Consequently, we need only consider such terms in the limit of large N.

By similar arguments one can show that all the clusters with an odd number of Hamiltonians vanish. Since all odd-order moments contain at least one such cluster, they must vanish asymptotically.

Even-order moments can be evaluated simply by counting the number of clusters of two H's since they form the dominating contribution to the values of the moments. For convenience we shall normalize the defining matrix elements of the Hamiltonian such that

$$\langle HH \rangle = 1, \tag{17}$$

i.e., the average trace of the square of H is taken to be unity. In counting the number of contributing terms, we should be careful about terms of the type $\langle \cdots ABCA \cdots \rangle$. As long as $m \gg k$, the single-particle states acted upon by the different Hamiltonians are likely to be different for the low-order moments. As a result, there is only a small probability that the single-particle states changed by the action of the first A will be changed by the intervening operators B and C in such a way that the action of the second A will not be able to restore these single-particle states back to the original ones prior to the action of the first A. For $m \gg k$, we can ignore this possibility since it can only happen to a small fraction of members in the entire ensemble. However, this will not be true if k is comparable with m. As we shall see later, this point is crucial in resolving the problem concerning the very different behaviour of eigenvalue distributions between different random matrix ensembles.

For $k \ll m$, the moment of order 2μ is equal to the number of different products of clusters of two that can be formed out of 2μ Hamiltonians. Hence

$$\overline{M}_{2\mu} = \frac{(2\mu)!}{2^\mu \mu!} = (2\mu - 1)!! \,, \tag{18}$$

where the bar over $M_{2\mu}$ indicates that it is an ensemble averaged value. Putting this result together with the fact that all the odd moments vanish, a Gaussian distribution is obtained.

What we have established so far is that the ensemble-averaged density distribution for Hamiltonians with particle rank $k \ll m$ is Gaussian. We must now go a step further and find out also whether most of the members in the ensemble have the same property. This can be done by checking the ensemble variance of the moments. The derivation is more involved and we shall only quote the results from Mon and French (1975). Since the first two moments serve only to set the origin and the scale factor, we start with the third moment. The general results for the variance with arbitrary k and m have been worked out for the low-order moments and we give here the results of ensemble variances for moments up to order six,

$$\overline{V_3}(m,k) = 6\binom{m}{k}^2 \binom{N}{k}^{-2} \left\{ 1 + 3\binom{m}{k}^2 \right\}, \tag{19a}$$

$$\overline{V_4}(m,k) = 8\binom{m}{k}^2 \binom{N}{k}^{-2} \left\{ 1 + 2\left[2\binom{m}{k} + \binom{m-k}{k} \right]^2 \right\}, \tag{19b}$$

$$\overline{V_5}(m,k) = 10\binom{m}{k}^2 \binom{N}{k}^{-2} \left\{ 1 + 15\left[\binom{m}{k} + \binom{m-k}{k} \right]^2 \right.$$
$$\left. + 5\binom{m}{k}^2 \left[2\binom{m}{k} + \binom{m-k}{k} \right]^2 \right\}, \tag{19c}$$

$$\overline{V_6}(m,k) = 12\binom{m}{k}^2 \binom{N}{k}^{-2} \left\{ 1 + 6\left[2\binom{m}{k} + 3\binom{m-k}{k} \right]^2 \right.$$
$$+ 3\left[5\binom{m}{k}^2 + 6\binom{m}{k}\binom{m-k}{k} + 3\binom{m-k}{k}^2 \right.$$
$$\left. \left. + \binom{m-k}{k}\binom{m-2k}{k} \right]^2 \right\}. \tag{19d}$$

We see that as long as $m \ll N$, the variances are small and the chance of finding a member that differs significantly from the ensemble average is negligible.

Let us return now to the more general case of a k-body Hamiltonian without the restriction $k \ll m$. In such cases Mon and French (1975) found, by working out all the possible combinations of clusters, that

$$\overline{M_4} = \binom{m}{k}^{-1} \left\{ 2\binom{m}{k} + \binom{m-k}{k} \right\}, \tag{20}$$

$$\overline{M_6} = 5 + \binom{m}{k}^{-1} \left\{ 3\binom{m-k}{k}\overline{M_4} + \frac{(3k)!}{(k!)^3}\binom{m}{k}^{-2}\binom{m}{3k} \right\}. \tag{21}$$

From eq. (20) we obtain an excess of -0.6 for the numerical example ($m = 6$, $k = 2$) described earlier. Unfortunately no general expression is available for all the moments for arbitrary m and k. The only exception is for $k = m$, i.e., many-body interactions. In such a case terms of the type $\langle \cdots ABCA \cdots \rangle$ do not contribute to the trace asymptotically and we obtain the well known semi-circular density distribution of eq. (11) given first by Wigner (1955).

The variation of the eigenvalue distribution as the maximum particle rank of the Hamiltonian is changed from $k = 2$ to $k = 7$ is demonstrated

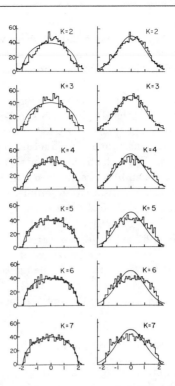

Fig. II-2. Level density distribution for seven-particle, 50-dimensional ensembles with the particle rank of the interaction varied from $k = 2$ to 7 (Wong and French, 1972). Each histogram is compared with a semicircular curve on the left and a Gaussian curve on the right. As $k \rightarrow m$, the level density distribution goes from Gaussian to semicircular.

by numerical simulation for an $m = 7$ system and the results are shown in Fig. II-2. For computational ease, there are only two active orbits and the seven-particle space, with $(J, T) = (7/2, 7/2)$, is only 50 dimensional. In spite of the small space, it is clearly demonstrated that, as k decreases from m, the density distribution rapidly changes from semicircular to Gaussian. This resolves one of the early puzzles of random matrix studies in that the GOE density distribution has a concave curve at low energies while the experimental curvature is convex, roughly of the form $\exp\sqrt{aE}$, and results calculated with the shell model agree with experiment. The question is understood once the dependence of the interaction on the particle rank is known.

II.7   Secular variation and fluctuations

What we have discussed so far is mainly concerned with the low-order moments that govern the secular variations of the state density with energy. The results are quite conclusive since the ensemble distributions of the moments are found to be concentrated in a narrow range as shown by the small ensemble variances. This means that, although we do not know the true Hamiltonian, it is likely that the ensemble-averaged results are representative of those of the true Hamiltonian. For an interaction dominated by parts with particle rank $k = m$, we expect the eigenvalue distribution to be semicircular. On the other hand, for a nuclear Hamiltonian with $k \leq 2$ a Gaussian distribution is expected as long as the number of active nucleons is much larger than 2.

In addition to smooth energy dependences, level densities and transition strengths display also short-range variations on the order of a few level spacings, generally known as fluctuations. There are several interesting questions concerning fluctuations that random matrix studies wish to address. First of all, one wants to know whether fluctuations in random matrix ensembles are the same as in experimental data. Second, are the fluctuations in different ensembles similar or different? Finally, and of more interest to statistical spectroscopy, we wish to know whether it is important to be able to calculate the fluctuations of nuclear observables.

In contrast to secular variations, fluctuations cannot be compared easily by inspection or in terms of moments. Statistical measures must be designed to "parametrize" the fluctuations so that quantitative studies can be made. As we shall see later, the conclusions concerning fluctuations depend to a large extent on the measures available.

II.8   Experimental situation

Let us start by first reviewing the experimental data. In order to study the distribution of energy levels, we need a large number of states within a narrow energy span. Furthermore, the energy interval involved must be small to ensure that the average level spacing D, as well as other possible underlying scaling parameters, remains unchanged for all the levels in the energy region. In addition, the nuclear Hamiltonian has certain exact symmetries, such as spin (J), parity (π), and isospin (T). Consequently, each state is characterized by a set of good quantum numbers (J^{π}, T). Mainly because of the lack of level repulsion, relationships between states with different good quantum numbers are quite different from those between states having the same ones. Hence, in addition to energy, complete (J^{π}, T) identification of a sequence of levels is also necessary for fluctuation studies.

Experimentally there are only two energy domains for which data are available. The first is the low-lying region starting from the ground state up to energies where the level density becomes too high and too many competing decay channels are open for clear spin and parity assignments to all the levels. For light nuclei, say $A \simeq 40$, the upper limit for this region is around 7 MeV and there are about 20 levels with 10 different (J^{π}, T) combinations in this range. As a result, only one to two spacings between levels of the same (J^{π}, T) in each nucleus are available for fluctuation studies. The situation is not too different for heavier nuclei except that the energy scale involved is smaller.

A second region is just above the neutron separation threshold, around 15 MeV in light nuclei and 7 MeV in heavy ones. The absorption of slow neutrons, up to a few keV in energy, in many nuclei is dominated by s-wave ($l = 0$) resonances. If the target nucleus, such as ^{166}Er, has a $J^\pi = 0^+$ ground state, the final states formed by capturing an s-wave neutron have $J^\pi = 1/2^+$ (since a neutron has intrinsic spin $1/2$). As a result, series up to $\approx$100 levels within an energy span of a few keV have been identified (Liou et al., 1972). The purity of the sequence depends in part on the strength of the p-wave ($l = 1$) resonances in the region, some of which may be strong enough to be detected and mistaken to be weak s-wave ones, and in part on the instrumental resolution which determines the minimum strength of a resonance before it can be detected. For ^{166}Er, it is believed that, out of a sequence of 109 levels, there is a possible admixture of four p-wave resonances and five possible s-wave ones undetected in the energy region. Unambiguous J^π identification for a level normally requires an angular distribution study. This is, however, too tedious to be carried out in a region of such high level density, especially for neutrons.

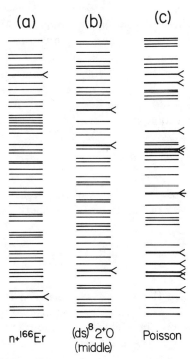

(a)   (b)   (c)

Fig. II-3. A section of 50 levels taken from (a) neutron resonance on an ^{166}Er target, (b) shell model eigenvalues for $(ds)^8$ $(J^\pi, T) = (2^+, 0)$ space, and (c) Poisson spectrum made of random numbers arranged in ascending order. Arrow heads mark spacings less than a quarter of the average in the region.

$n + ^{166}$Er $(ds)^8 2^+ 0$ Poisson
                      (middle)

Proton resonance data are also available but only at somewhat lower excitation energies (Bilpuch et al., 1976). The level density in this region is much lower and, as a result, much larger energy intervals are needed before one can have enough levels for a statistical study. It is therefore necessary to "unfold" the spectrum to take away the variation of D as a function of energy. This is in general a fairly reliable procedure except when there is the complication of *doorway* states, which introduce local variations in the level density and transition strengths that are longer in range than the few level spacings typical of fluctuations but much shorter than those characterizing secular variations. Since the presence of a doorway is in many ways equivalent to the introduction of a new degree of freedom that is local to the energy region, it becomes a problem for standard random matrix studies. On the other hand, because of the lower level density and the fact that protons are involved, better identification of the orbital angular momentum l can be achieved with proton resonances.

Most of the experimental measurements have gone through a certain amount of screening using statistical tests to assess the quality of the data. The amount of *adjustments* made to the data is not insignificant even for ^{166}Er, one of the best experimental sequences available. As we shall see later, these "tests" are by no means unrelated to the statistical measures used to study the fluctuation properties of the spectrum. Caution is therefore called for in many cases. Much more rigorous statistical analyses are possible (Coceva and Stefanon, 1979) but have not been applied in most of the experimental data reductions.

For the neutron resonance data on ^{166}Er, the density of states is constant for the entire 109 level span: the spectrum, however, departs noticeably from a smooth, *i.e.*, evenly spaced, one as can be seen from a 50 level section of it shown in Fig. II-3a. To *see* the fluctuations, we can plot the distribution, in the form of a histogram, of the nearest neighbour spacing

$$S_i = (E_{i+1} - E_i), \tag{22}$$

where E_i is the energy of the level i. The result for the entire 108 spacings is shown in Fig. II-4a. The smooth curve given by

$$P(S) = \frac{\pi S}{2D^2} \exp\left(-\frac{\pi S^2}{4D^2}\right), \qquad S \geq 0, \tag{23}$$

is called the Wigner surmise (von Neumann and Wigner, 1929, Wigner, 1967) derived using level repulsion arguments. The form given in eq. (23) was shown by Mehta and Gaudin (1960) to be an extremely good approximation to the exact results.

In contrast, it is interesting to compare with the case of replacing the energy levels E_i by a set of random numbers arranged in ascending order. The levels in such a *Poisson* spectrum are completely uncorrelated and, as a result, there is no repulsion between them. The distribution of S, instead of being given by eq. (23), changes to $P(S) = D^{-1} \exp(-S/D)$.

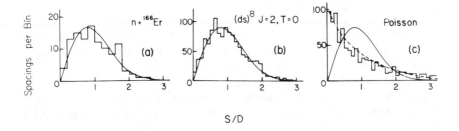

Fig. II-4.   Nearest neighbour spacing distributions for the three cases
of Fig. II-3. The smooth curves are based on the Wigner surmise of
eq. (23), and the dashed curve in (c) is the function $D^{-1}\exp(-S/D)$.

II.9   Numerical simulation of data

Partly because of the scarcity of good quality data, numerical modelling
has become an integral part of fluctuation studies. By taking a suitable
Hamiltonian, the energy levels of a nucleus can be calculated, say, using the
shell model. The calculated results have been shown to be quite successful
in explaining many of the observed nuclear properties at low energies. The
discrepancies found can usually be attributed to the small size of the space
used and the shortcomings of the Hamiltonian. Our interest here lies solely
in using the nuclear shell model as a tool for numerical simulation of "data"
in as realistic a manner as possible.

With shell model eigenvalues there is no longer any question as to
the purity of the sequence. However, since the eigenvalue problem must
be solved numerically, there is a very severe restriction on the maximum
size of the matrix one can handle. As a result, it is not possible to obtain
a long sequence within a region of constant level density. This difficulty
is usually circumvented by *unfolding*, that is, transforming the eigenvalue
distribution into a constant one by assuming some simple functional form,
usually parametrized in terms of low-order moments of the distribution,
for the secular energy variation of the level density. So long as there is
a separation of the role of low- and high-order moments, the removal of
low-order ones by unfolding cannot affect the fluctuation behaviour of the
eigenvalue distribution.

As we see from Figs. II-3 and II-4, the shell model spectrum and its
nearest neighbour spacing distribution are very similar to those observed in
the experimental data. This is also true for the other fluctuation measures
to be described later.

II.10 Nuclear-table ensemble

As mentioned earlier, the spin, parity and isospin of the states in the ground state region of many nuclei are known. From these data, we can extract a few spacings between states of the same (J^π, T) in each nucleus. Together they constitute a large sample of experimental spacing data with well identified energy, spin, parity, and isospin. The only trouble is that the underlying scaling parameter D varies from nucleus to nucleus. As a result, we cannot make statistical studies unless we have a model for the variation of D as a function of the nucleon number A. This is also a kind of "unfolding", except that instead of the energy dependence, we are trying to remove the A dependence of D.

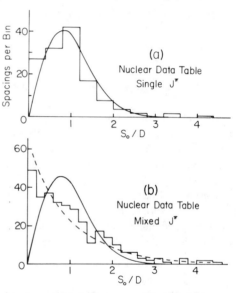

Fig. II-5. Nearest neighbour spacing distributions in the ground state domain: (a) for spacings between states of the same (J^π, T), and (b) for spacings regardless of (J^π, T). See also Figs. II-3.

In general, the average spacing in nuclei decreases with increasing nucleon number A. Local departures from this simple rule exist for the very light ones, those near closed shells and "collective" nuclei. If data from these "special" cases are excluded from the sampling, a $D \propto A^{-1}$ rule is found to be adequate for scaling all the remaining spacings. Such a collection of experimental data is called the nuclear-table ensemble (Brody *et al.*, 1976). The nearest neighbour spacing distribution of the ensemble is shown in Fig. II-5a. The agreement with the Wigner surmise is quite good. In contrast, if the ground state spacing, *i.e.*, the spacing between the ground state and the first excited state regardless of the (J^π, T) values (they are different from each other in most cases), the distribution, after scaling by the same $D \propto A^{-1}$ rule, is close to a Poisson one as shown in Fig. II-5b.

This result is significant for several reasons. The first is that nearest

neighbour spacings between states of different (J^π, T) are found to be uncorrelated, as expected from the absence of level repulsion and confirmed
by shell model simulations. Experimentally the ground state region is one
of the few places that such observations can be made. The second is that,
as far as the nearest neighbour spacing distribution is concerned, the fluctuations of energy level positions are the same in the ground state domain
as at higher energies. This is not necessarily an expected result, since there
may well be good arguments that the ground state of a quantum mechanical system is special by virtue of being the lower bound of an otherwise
infinite spectrum. However, the result of the nuclear-table ensemble shows
that the ground state is just another state in the spectrum having the same
fluctuation properties as the rest.

Unfortunately there are not sufficient data for testing with other statistical measures. Random matrix ensembles and shell model results are
not of great help since the energy variation of D is most rapid at the ends
of the spectrum. Furthermore, the unfolding procedures, designed for using in the middle of a spectrum, are not reliable in a region of low state
density. As a result, it is not easy to make any meaningful statement for
the ground state region.

Some support for this conclusion concerning the ground state can also
be obtained in the following way. One of the characteristics of energy levels
is their rigidity, *i.e.*, in an ensemble a particular level does not move very
much from one member to another. Numerical calculations, confirmed by
analytical studies, show that it is unlikely for a level to move more than a
distance D away from its average position in the ensemble. For example,
using the centroid of each spectrum as the reference point, the ensemble
average position of the level r will be at position rD. In a given member of
the ensemble, the level is most likely to be found within an interval $(r\pm1)D$,
where the value of D for each member is obtained from the levels in that
member to avoid finite size effects (Wong and French, 1972). One way
to see whether the same fluctuation pattern extends to the ground state
region is to check whether spectrum rigidity also extends to the ground
state.

In Fig. II-6, the distributions of ground state location as measured from
the centroid of each member are plotted for the 294-dimensional GOE and
TBRE. To avoid uncertainties in unfolding, the width of the eigenvalue
distribution is used instead of D as the energy unit. For GOE, the distribution is sharply peaked with a width less than one average local spacing.
For TBRE the spread is much larger (width $\approx 1.3D$) but still characterized
by a standard deviation of around one level spacing in the ground state region. The larger spread in the ensemble distribution for TBRE compared
with GOE is mainly due to the far smaller numbers of degrees of freedom
in the defining matrix elements of the system, and the same phenomenon
is also observed in the distributions of other measures.

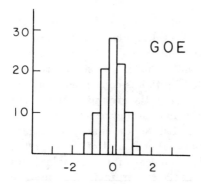
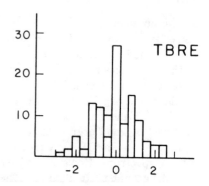

Fig. II-6. Distributions of the ground state positions for members of an ensemble. The horizontal axis is in units of the width of the eigenvalue distribution.

II.11 Fluctuation measures

We have seen that the nearest neighbour spacing distribution is a way to display the fluctuations of energy level positions and eigenvalues. In analogy with eq. (22), we can define spacing of order k as

$$S_i^k = (E_{i+k+1} - E_i), \tag{24}$$

i.e., the spacing distance between two levels with k levels in between. As k increases, the distribution of S^k in units of D, the average of S, approaches a Gaussian distribution centered at $k+1$ for both GOE and TBRE. This means that levels far away from each other are essentially uncorrelated. In fact, the deviation from Gaussian becomes indistinguishable for $k \geq 8$, as shown in Fig. II-7. Experimental data also seem to follow the same trend; however, there are usually not enough levels to make a meaningful plot for higher k. Similarly it is not useful to display the higher-order spacing distribution for shell model results as it becomes more and more susceptible to errors in unfolding.

In order to be more quantitative, it is necessary to characterize the fluctuations in terms of statistical measures. For a statistic to be useful, it must have a narrow ensemble distribution, preferably with known ensemble variance. In this way, any departure from the ensemble average value can be compared with the square root of the variance to determine whether the deviation is a significant one. In addition, the measure must be easy to evaluate and as insensitive to impurities in the sequence as possible.

Departures from the Wigner surmise for the nearest neighbour spacing distribution can be characterized in terms of the level repulsion parameter ω (Brody, 1973). The statistic is useful in many circumstances; however,

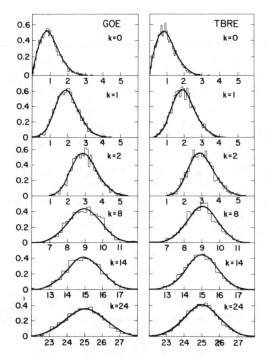

Fig. II-7. Histograms of the spacing distribution of order k for different k values. The smooth curves are theoretical GOE results for $k = 0, 2$, and Gaussian for higher k values.

since its ensemble behaviour is not well known, we shall not deal with it here. The same is true for the Λ statistic (Monahan and Rosenzweig, 1972) which is also based on the nearest neighbour spacing distribution.

The distribution of level spacings S^k can be characterized in term of the width σ_k, the square root of the variance σ_k^2. For $k = 0$, a fairly accurate value of σ_0 can be worked out from eq. (23). Other results for GOE up to $k = 8$ have been obtained by Mehta (as reported by Bohigas and Giannoni, 1975) and are listed in Table II-1. The ensemble widths of the distributions of σ_k are not available analytically. The values calculated from the 294-dimensional GOE and TBRE are on the order of 0.1 for low k and rise to around 0.15 for $k \approx 10$.

Instead of σ_k, we can also use the correlation coefficient $C(r)$ between two nearest neighbour spacings S_i separated by r spacings in between,

$$C(r) = \frac{n}{n-r-1} \frac{\sum_i (S_i - D)(S_{i+r+1} - D)}{\sum_i (S_i - D)^2} , \qquad (25)$$

to characterize the distribution of a run of n spacings. The relation between the two sets of quantities $C(r)$ and σ_k is given by (Garrison, 1964, Bohigas and Giannoni, 1975)

$$C(r) = \frac{1}{2\sigma_0^2} \left\{ \sigma_{r+1}^2 - 2\sigma_r^2 + \sigma_{r-1}^2 \right\} . \qquad (26)$$

For $r = 0$, $C(0) = -0.271$. The negative sign means that a small spacing is likely to be followed by a large one, a consequence of the level repulsion and spectral rigidity phenomena. Since $C(r)$ and σ_k give the same kind of information, only the value of $C(0)$ is listed in Table II-1 in conformity with the general practice in the literature.

Another way to measure fluctuations is to find, in an interval rD, how much the observed number of levels differs from r. The value of r should be small enough so that, for a sequence of n levels, one can take as many independent spans of rD as possible. On the other hand, r must be large enough compared with unity so that the interval will have enough levels for the statistic to be meaningful. Since D is fixed by the energy range covered by the n levels, the average deviation from the expected number is zero. The variance, the average of the square of the deviation, has the value

$$\Sigma^2(r) = \frac{2}{\pi^2} \left\{ \ln(2\pi r) + 0.5772\cdots + 1 - \frac{\pi^2}{8} \right\} + O\left(\frac{1}{\pi^2 r}\right), \qquad (27)$$

when averaged over the ensemble. In Table II-1, the square root of $\Sigma^2(r)$ for $r = 10$ is given for various ensembles, shell model results, and experimental data. A more exact value of $\Sigma^2(10)$ than the value 0.91 given in eq. (27) is 0.95 obtained for GOE (Brody *et al.*, 1981).

To test the purity of an experimental sequence, Dyson introduced the F-statistic (reported in Liou *et al.*, 1972) based on the expected correlation between spacings in an interval of length L. The correlation function used is

$$f(x) = \begin{cases} \frac{1}{2} \ln\left(\frac{1+\sqrt{1-x^2}}{1-\sqrt{1-x^2}}\right), & \text{for } |x| < 1, \\ 0, & \text{otherwise.} \end{cases}$$

At level i, the value of F is

$$F_i = \sum_{j \neq i} f\left(\frac{E_j - E_i}{L}\right). \qquad (28)$$

For $L = rD$, the expectation value of F is

$$\langle F \rangle = r\pi - \ln r\pi - 0.656,$$

with a variance of $\sigma^2(F) = \ln r\pi$.

At a spurious level, *i.e.*, a level that is not a member of the sequence, we have $\langle F \rangle = r\pi$. Thus, for $r \approx 10$, the presence of an impurity can be identified. For a pure sequence, the distribution of F is Gaussian: departure from a Gaussian distribution for F therefore can also serve as an indication of the presence of impurities. However, Monte Carlo tests have shown that the distribution of F will return to be Gaussian, albeit with larger variance, once the impurity is more than about 10%. Hence, a Gaussian distribution of F alone is not adequate evidence for a pure sequence.

Two other commonly used statistics, Δ_3 and Q, were introduced by Dyson and Mehta (1963). Δ_3 measures the fluctuations by evaluating the deviation of levels from a smooth spectrum in terms of a staircase function $F(x)$ with energy (in the form E/L) as the variable. At the position of each energy level, the value of $F(x)$ goes up by unity. For a sequence of n levels,

$$\Delta_3(n) = \frac{d}{2L} \min_{A,B} \int_{x-L}^{x+L} \{F(x') - Ax' - B\}^2 \, dx, \qquad (29)$$

where $2L = nD$. For GOE, the ensemble average value is

$$\overline{\Delta_3} = \frac{1}{\pi^2}\{\ln(n) - 0.0687\}, \qquad (30)$$

whereas for a completely random sequence the value is $n/15$. The original intention of Δ_3 was again to detect the presence of impurity in an experimental sequence. Thus, if n is large, it is easy to identify whether the sequence of levels contains a significant number of impurities. Furthermore, Δ_3 has the additional advantage that the ensemble variance is independent of n to a first approximation. Recent improvements in this measure make it even more attractive as a statistic for fluctuation studies (Haq, Pandey, and Bohigas, 1982).

The Q statistic measures the short-range correlations between levels involving a sum with weighting factor $\ln 2\pi |E_i - E_j|/D$. Counter terms are added to moderate the strong variations of Q due to the variation of the position of a particular level in different members of the ensemble. The form of the measure is too long to be given here. For GOE the ensemble average and variance for an interval of n levels are

$$\overline{Q(n)} = n\left\{0.365 - \frac{1}{r\pi^2}\right\}, \qquad \overline{\sigma^2(n)} = n\left\{0.266 - O\left(\frac{1}{r\pi^2}\right)\right\}, \qquad (31)$$

where r should be taken to be between about 2 and 4.

II.12   Results of fluctuation studies

We have used shell model results and ensembles of random matrices to supplement experimental data for the study of energy level fluctuations in nuclei. Although we have been concerned mainly with TBRE and GOE, several other ensembles are available (Wigner, 1972). However, for an ensemble to be useful, it must be physically relevant and mathematically convenient. Ideally, as we learn more about the nuclear system, it may be possible to construct ensembles for the nuclear system that are more restrictive than TBRE so that we may average over an even smaller set of "reasonable" Hamiltonians.

There are also two other questions the answers to which are not easily forthcoming. The first one is ergodicity (Pandey, 1979). Although we do not have time development here as in ordinary statistical mechanics, we need nevertheless to ensure that all the members of the ensemble are equally representative of what may be obtained with the true Hamiltonian. The second question is whether the "reasonable" Hamiltonians dominate in the ensemble. If this is not true the ensemble average may not be of interest regardless of the narrowness of the ensemble distribution. An example of this can be found in the case of the level density distribution. GOE is a very general ensemble and it contains TBRE as a small part of it. Even though the low-order energy moments in GOE have very narrow ensemble distribution, they are not representative of the eigenvalue distributions for two-body operators. As we have shown earlier, the GOE results for eigenvalue distributions are actually quite different from those for TBRE and from experimental data.

The measures described in the preceding section, as well as several others not given here, have been applied to a variety of ensemble averages, shell model eigenvalues, and experimental spectra. A sample of the results obtained is given in Table II-1. It is seen that as far as the measures are concerned, the fluctuations are the same in all cases. Except for those indicated by an asterisk, all the values fall within one standard deviation of the results expected from GOE. The number of cases outside the one standard deviation criterion is within that expected for a "random" sample of cases. In a more complete survey (Brody *et al.*, 1981), a greater number of deviations is found. These are inevitably associated with particular pieces of data. In other words, if a spectrum does not fit the GOE value for one of the measures, there is a high probability that it also fails to meet the GOE expectations for other measures. One of the possible conclusions is that the datum itself is faulty and this is often corroborated by experimental evidence, such as the existence of large numbers of possible missing and spurious levels.

On the other hand, one should not be surprised to see a high degree of correlation among the results for different measures. In addition to exact relations between them, such as that given by eq. (26), all the known measures are essentially given by a two-point or covariance function (Brody *et al.*, 1981),

$$S^\rho(x,y) = \overline{\rho(x)\rho(y)} - \overline{\rho(x)}\,\overline{\rho(y)}, \tag{32}$$

where $\rho(x)$ is the state density distribution and the overbar indicates ensemble averaging. In other words, all the available fluctuation measures depend essentially on the correlations between two points in the spectrum. In principle, higher-point measures depending on the correlations between several levels can be designed, but none is known to have been applied.

Table II-1 Fluctuation statistics for different cases[†]

Case	n	$C(0)$	σ_0	σ_3	σ_6	$\sigma(F)$	$\Sigma(10)$	Δ_3	Q
GOE(∞)	100	−0.27	0.53	0.75	0.82	1.9	0.95	0.46	34.0
GOE(294)	100	−0.27	0.54	0.75	0.84	1.9	0.88	0.46	34.2
TBRE(294)	100	−0.26	0.52	0.73	0.80	2.0	0.96	0.46	34.5
$(ds)^8 2^+ 0(1206)$	120	−0.30	0.50	0.67	0.68	1.3[*]	0.81	0.33(0.48)	29.7(41)
$(ds)^{12} 0^+ 0(839)$	84	−0.24	0.52	0.87	0.97	1.9	0.95	0.51(0.44)	34.7(29)
^{166}Er	109	−0.22	0.53	0.76	0.81	1.8	0.96	0.46(0.47)	30(36)
^{168}Er	50	−0.29	0.50	0.73	0.65	1.9	0.78	0.29(0.39)	15(16)
^{152}Sm	70	−0.26	0.52	0.94[*]	0.82	1.7	0.85	0.40(0.42)	25(23)
^{232}Th	178	−0.19	0.54	0.83	0.86	2.2	1.05	0.39[*](0.51)	69[*](60)
^{238}U	146	−0.24	0.50	0.60	0.62	1.7	0.80	0.42(0.49)	32[*](49)
^{172}Yb	55	−0.24	0.55	0.88[*]	1.00[*]	1.7	1.04	0.41(0.40)	18(18)

[†]The results for GOE(∞) are analytical; all others are obtained numerically. For Δ_3 and Q the theoretical values depend on n and are given in brackets; for all others the theoretical values are given by the GOE(∞) case. The ensemble width of the 294-dimensional TBRE and GOE is 0.8 for $C(0)$ and 0.3 for $\sigma(F)$. Asterisks indicate that the value deviates from the ensemble average value by more than one standard deviation.

As far as the available measures can tell, fluctuations are the same for both GOE and TBRE in spite of the important difference in their physical basis and in the values of their low-order moments. The fact that the results of these measures are also the same for the experimental data shows that we are studying something that is physical, and not merely a construct of our ensembles. Very few assumptions are made in defining these ensembles. The main one is that the system is a time-reversal invariant, quantum mechanical system of fermions interacting with a k-body force. From the similarity between GOE and TBRE, we have eliminated any dependence of fluctuations on the particle rank of the interaction. We are therefore led to the inevitable conclusion that fluctuations found in nuclear spectra are general properties of quantum mechanical systems.

There seemed to be some hope of using fluctuation measures to distinguish between different systems, *e.g.*, between two-body and many-body interactions. Certainly this is not possible unless a set of measures independent of the presently available ones can be found capable of making a distinction between different ensembles. However, this seems to be unlikely.

This failure is not necessarily a disappointment. What we have seen is perhaps an indication that fluctuations in eigenvalue distributions are general quantum mechanical phenomena not limited to nuclear spectra alone. If this is true, it is of interest to find out how broad the universality is. Our interest here, however, is quite different. If the behaviour of fluctuations is only a general quantum mechanical property, we cannot expect to learn anything about the nuclear system itself from it. As a result, we can concentrate our efforts on the secular variations governed by the low-order moments of the distribution.

II.13   Distribution of excitation strengths

In addition to energy level positions, the fluctuations in excitation strengths have also been well studied. An excitation connects an initial state $|E\rangle$ at energy E to a final state $|E'\rangle$ at energy E'. The strength $R(E', E)$ can be defined as

$$R(E', E) \equiv |\langle E'|\hat{O}|E\rangle|^2 = \langle E|\hat{O}^\dagger|E'\rangle\langle E'|\hat{O}|E\rangle, \qquad (33)$$

where $\hat{O}$ is the excitation operator. Depending on the nature of $\hat{O}$, the two states involved may be in the same space, as *e.g.*, for an E2 transition between states of the same (J^π, T); or they may be in two spaces with different particle number, as for a one-nucleon transfer reaction in which the initial state is in a nucleus with A nucleons and the final state in one with $(A \pm 1)$ nucleons. In either case, the strength is a function of both the initial and final state.

In the same way as we have done for eigenvalue distributions, we can separate the energy dependence of $R(E', E)$ into two parts; a long-range part that changes only slowly with E and E', and a fluctuating part that changes rapidly from one pair of states to another. It is well known that the fluctuations in excitation strength follow a Porter-Thomas distribution (Lynn, 1968), and we shall give below a brief argument for this result.

For simplicity we assume that $|E\rangle$ and $|E'\rangle$ are in the same space, and we let $\{\Phi_i\}$ be a complete set of orthonormal basis states in the space. Regardless of how we choose the basis, so long it is not trivially related to the eigenvectors, we can express $|E\rangle$ as a linear combination of the basis states,

$$|E\rangle = \sum_i C_i(E)|\Phi_i\rangle. \qquad (34)$$

For simplicity, we can take the expansion coefficients $C_i(E)$ to be real. Except for the normalization condition,

$$\sum_i C_i(E)^2 = 1, \qquad (35)$$

we can regard the expansion coefficients as essentially independent of each other. This is the main assumption used to derive the Porter-Thomas distribution, but no satisfactory proof of the assumption is known. However, all the available indications seem to confirm its correctness especially in large dimensional spaces.

With this assumption, we can proceed to derive the distribution of excitation matrix elements. In terms of the basis representation,

$$\langle E'|\hat{O}|E\rangle = \sum_{ij} C_i(E') \, C_j(E) \, \langle \Phi_i|\hat{O}|\Phi_j\rangle. \tag{36}$$

Furthermore, since the eigenvectors are orthonormal,

$$\sum_i C_i(E)C_i(E') = \delta_{E,E'} \, .$$

The average of $C_i(E)$ must therefore be zero and, as a result, each term in the sum in eq. (36) must also be zero on the average. The value of $\langle \Phi_i|\hat{O}|\Phi_j\rangle$ depends on the choice of the basis representation. However, for strength distribution studies, neither the value nor the exact form of the distribution of the matrix elements is of any importance to the following arguments. For simplicity, we take $\langle \Phi_i|\hat{O}|\Phi_j\rangle$ to be Gaussian random variables: the same final result can be obtained even if they are all assumed to be equal to each other.

In any case, because of the independence of the $C_i(E)$ from each other, each of the terms in the sum in eq. (36) can be regarded as an independent zero-centered random variable. Regardless of the distribution of each one of these random variables, the distribution of $y \equiv \langle E'|\hat{O}|E\rangle$ is a zero-centered Gaussian,

$$P(y)\, dy = \frac{1}{\sqrt{2\pi\sigma^2}} \exp\left(-\frac{y^2}{2\sigma^2}\right) dy, \tag{37}$$

in virtue of the central limit theorem (Cramer, 1946).

The excitation strength is the square of the matrix element. Let

$$x \equiv y^2 = R(E', E).$$

The distribution of excitation strength is then

$$p(x)\, dx = \frac{1}{\sqrt{2\pi\sigma^2}} \exp\left(-\frac{x}{2\sigma^2}\right) \frac{dx}{\sqrt{x}} = \frac{1}{\sqrt{2\pi x \bar{x}}} \exp\left(-\frac{x}{2\bar{x}}\right) dx. \tag{38}$$

In the final result we have replaced σ^2 by $\bar{x}$ to put it into the familiar Porter-Thomas form in terms of a chi-square distribution of one degree of freedom. In going from eq. (37) to eq. (38), note that $-\infty \geq y \geq +\infty$ whereas $0 \geq x \geq +\infty$, and both distributions are normalized to unity.

In eq. (38), $\bar{x}$ is the variance of the distribution of excitation matrix elements in eq. (37). Since y is zero centered, we have

$$\bar{x} \equiv \sigma^2 = \int y^2 \, p(y) \, dy = \int x \, p(x) \, dx = \overline{R(E', E)}. \tag{39}$$

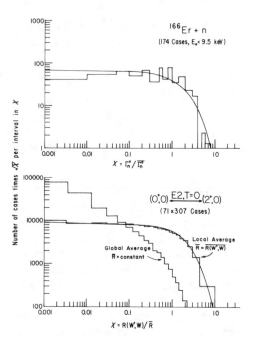

Fig. II-8. Histogram of the reduced width for slow neutron resonances on a ^{166}Er target (Liou, *et al.*, 1972). The result is in good agreement with the Porter-Thomas distribution given by the smooth curve. The same agreement is also obtained for the shell model E2 excitation strength if the local average is used instead of the global average for the value of $\bar{x}$ in eq. (38).

Hence we find that $\bar{x}$, the scale factor for the distribution of excitation strengths, is the average value of the excitation strength itself. This is a very reasonable result which we could have anticipated. However, we have used a constant scale factor throughout the space by implicitly assuming that the distribution of x has the same variance everywhere. This is obviously incorrect if there are any secular variations in the excitation strengths with respect to E and E'. As we shall see in the next chapter, unless the excitation operator $\hat{O}$ is totally uncorrelated with the Hamiltonian, we have no reason to expect that a constant scale factor for the fluctuations of excitation strengths is valid for the whole space. In general, we must take the locally averaged value of $R(E', E)$ as the scale factor $\bar{x}$ in eq. (38).

Fig. II-8 illustrates the situation with a shell model example. When a global scale factor is used, the distribution of strengths is far from that given by a Porter-Thomas distribution. On the other hand, the same set of excitation strengths scaled by the locally averaged value follows the Porter-Thomas distribution to a high confidence limit. The physical reason behind this is that only the fluctuations of the excitation strength are described by the Porter-Thomas distribution: any secular variations of the local average are not part of the fluctuations and must therefore be removed before we can obtain the expected distribution.

Experimentally one does not usually need to consider the variation of $R(E', E)$ since the data normally do not cover a sufficiently wide energy domain. However this is not strictly correct either. For example, there are occasions where the data include the influence of a doorway state. In such a case, the general magnitudes of the excitation strengths undergo a smooth local variation across the span of the doorway. Such a variation cannot be regarded as part of the fluctuations especially since the energy span of the doorway is much wider than a few states. By fluctuations, we usually mean the rapid variations of the magnitude in the region of a few states. Unless we have a theory for the energy variation of the average strength over a doorway, we may have the practical difficulty of finding the local average in such a situation. Because of such difficulties, there is occasional confusion in the literature whether a violation of the Porter-Thomas distribution is found. There is also some doubt whether, in the presence of a doorway or other similar local perturbation, the distribution of excitation strengths will be Porter-Thomas. Unless a proper subtraction of the variation of the local average is removed, it is not possible to give a meaningful answer to this question.

II.14   Summary and conclusion

We have seen that, without making any recourse to the nuclear Hamiltonian, it is possible to account for the fluctuations in both eigenvalue and excitation strength distributions using statistical arguments alone. Except for the slow variations of the underlying scale factors, such fluctuations cannot therefore have any possible connections with nuclear physics. This is fortunate since such secular behaviour is related to the low-order moments of the distributions, and powerful methods can be developed to calculate these moments.

This is especially important for nuclear physics since the nature of the interaction is such that, in general, extremely high-dimensional spaces are required for realistic calculations. Shell model and other microscopic techniques fail here since, in order for the calculations to be tractable, severe truncation of the space is required. In statistical spectroscopy we attempt, instead of reducing the active space, to evaluate low-order moments that describe the smooth behaviour of the distribution of the various quantities of interest. In this way, it is hoped that new insight into the problems may be obtained.

Chapter III

Energy Levels and Excitation Strengths

From the results of random matrix studies made in the last chapter, we find
that fluctuations in eigenvalue distributions are essentially independent of
the system from which the eigenvalues are derived. Hence we cannot expect
to learn anything about the structure of a nucleus from such studies. Since
fluctuations are determined by the high-order moments, we can conclude
that little information of interest to us is contained in these moments. From
now on, we shall therefore concentrate instead on the low-order moments
that characterize the smooth variation of the distributions and study their
connections with the nuclear system.

III.1 Distribution of eigenvalues

Let us start with the distribution of eigenvalues, also referred to as the
density function or density of states. For a system with m particles in
a space consisting of N single-particle states, it was shown in Chapter II
that, in the limit $N \gg m$, the eigenvalue distribution is Gaussian for a
Hamiltonian with low ($\ll m$) particle ranks. The limiting conditions are
not always satisfied in realistic situations, but we should be close to fulfill
them, and the eigenvalue distributions are expected to be approximately
Gaussian in general.

Given a set of moments defining a distribution that is nearly Gaussian,
we must find a way to *realize* the distribution itself, *i.e.*, to reconstruct the
distribution from the given moments. In other words, we wish to find a distribution having the same moments as the given set. If only an incomplete
set of moments is available, there is some ambiguity in reconstructing the
distribution and a model is required. For a nearly Gaussian distribution,
the most direct method is to use the Gram-Charlier series (Cramer, 1946),

$$\rho(x) = \rho_G(x) \sum_{\nu=0}^{\infty} \frac{S_\nu}{\nu!} He_\nu(x), \qquad (1)$$

where $x = (E - C)/\sigma$ and $\rho_G(x)$ is the Gaussian function defined in eq. (II-
9). The expression is exact if the complete set of moments is known.

However, this is not usually possible in practical cases in which the upper limit of the summation is governed by the highest-order moment available.

The Hermite polynomial $He_\nu(x)$ is given by

$$He_\nu(x) = (-1)^\nu \, e^{\frac{1}{2}x^2} \, \frac{\partial^\nu}{\partial x^\nu} \, e^{-\frac{1}{2}x^2}. \qquad (2)$$

The lowest orders have the explicit forms:

$$
\begin{aligned}
&He_0(x) = 1, && He_1(x) = x, \\
&He_2(x) = x^2 - 1, && He_3(x) = x^3 - 3x, \\
&He_4(x) = x^4 - 6x^2 + 3, && He_5(x) = x^5 - 10x^3 + 15x, \\
&He_6(x) = x^6 - 15x^4 + 45x^2 - 15.
\end{aligned}
\qquad (3)
$$

Since the Hermite polynomials form an orthogonal set,

$$\int_{-\infty}^{+\infty} He_\mu(x) \, He_\nu(x) \, \rho_G(x) \, dx = \mu! \, \delta_{\mu\nu}, \qquad (4)$$

the structure factors S_ν in eq. (1) are given by

$$S_\nu = \int_{-\infty}^{+\infty} \rho(x) He_\nu(x) \, dx. \qquad (5)$$

They are functions of the moments of $\rho(x)$ and can be used instead of the moments as the defining quantities for the distribution. For a distribution normalized to unity, and with x centered and measured in units of its width, i.e., $x = (E - C)/\sigma$, we have $S_0 = 1$ and $S_1 = S_2 = 0$. If all the moments up to order μ are known, we can construct a distribution $\rho^\mu(x)$ by a Gram-Charlier expansion in terms of Hermite polynomial up to order μ,

$$\rho^\mu(x) = \rho_G(x) \sum_{\nu=0}^{\mu} \frac{S_\nu}{\nu!} \, He_\nu(x). \qquad (6)$$

By inspection, it is obvious that $\rho^\mu(x)$ has the same moments as $\rho(x)$ up to order μ.

Instead of a Gram-Charlier series, an Edgeworth series (Cramer, 1946) expansion,

$$
\begin{aligned}
\rho(x) = \rho_G(x) \Bigg\{ 1 &+ \frac{\kappa_3}{3!} He_3(x) + \frac{\kappa_4}{4!} He_4(x) + \frac{\kappa_5}{5!} He_5(x) + \frac{10\kappa_3^2 + \kappa_6}{6!} He_6(x) \\
&+ \frac{35\kappa_3\kappa_4}{7!} He_7(x) + \frac{35\kappa_4^2 + 56\kappa_3\kappa_5}{8!} He_8(x) \\
&+ \frac{280\kappa_3^2}{9!} He_9(x) + \frac{2100\kappa_3^2\kappa_4}{10!} He_{10}(x) + \cdots \Bigg\},
\end{aligned}
\qquad (7)
$$

may also be used. In principle, the Edgeworth series is superior (Feller, 1971) since it is an expansion in terms of cumulants κ_μ rather than moments. As we have seen in Chapter II, the cumulants above order 2 vanish for a Gaussian distribution. Consequently, we expect that the expansion of a nearly Gaussian distribution converges more rapidly if carried out in terms of cumulants. In practice, however, there is little difference between the two methods if the expansions are restricted to the few lowest orders. This comes simply from the fact that, for the few lowest orders, the Hermite polynomial moments and cumulants are essentially identical, as can be seen by comparing eqs. (II-8) and (3). For expansions to higher orders there may well be a significant difference but such high-order expansions are not of general interest in statistical spectroscopy.

III.2   Partitioning of the space and eigenvalue distributions

The space of N single-particle states can be divided into several subspaces. The distribution of strengths in the entire space is then the sum over the distributions of the strengths $\rho_\alpha(x_\alpha)$ in each subspace. If the number of states in subspace α is d_α and each $\rho_\alpha(x_\alpha)$ is normalized to unity, we have

$$\rho(E) = \frac{1}{d} \sum_\alpha d_\alpha \, \rho_\alpha(E), \tag{8a}$$

with

$$d = \sum_\alpha d_\alpha . \tag{8b}$$

It is convenient for later to define also $I(x)$, the state density normalized to d instead of unity,

$$I(x) \equiv d \cdot \rho(x) . \tag{9}$$

Eqs. (8) can then be rewritten in the form

$$I(E) = \sum_\alpha I_\alpha(E), \tag{8c}$$

with the function $I_\alpha(E)$ in subspace α normalized to d_α.

The distribution of the strengths in each subspace can also be expanded in terms of a Gram-Charlier series in the same form as given by eq. (6), and eq. (8a) then becomes

$$\rho(x) = \frac{1}{d} \sum_\alpha d_\alpha \, \rho_G(x_\alpha) \sum_\nu \frac{S_\nu(\alpha)}{\nu!} He_\nu(x_\alpha), \tag{10}$$

where $x_\alpha = (E - C_\alpha)/\sigma_\alpha$, i.e., the subspace distribution centroid C_α is used as the origin and the width σ_α as the scale for the variable x. Each subspace has its own set of structure factors $S_\nu(\alpha)$, and the structure factors of the complete distribution can be expressed as a function of these. For example, in terms of moments the explicit relations for the lowest two orders are given by

$$C = \frac{1}{d} \sum_\alpha d_\alpha C_\alpha, \quad \text{and} \quad \sigma^2 = \frac{1}{d} \sum_\alpha d_\alpha \left(\sigma_\alpha^2 + C_\alpha^2 \right) - C^2. \quad (11)$$

Similar relationships can be derived between the higher moments.

On partitioning the space, the number of inputs to the distribution is increased since each subspace distribution has its own set of defining moments or structure factors. As a result, we can in principle extract more information from eq. (10) than from eq. (6). For example, from eq. (10) we can find out the relative contributions from different subspaces at a given energy E, and this is impossible if we have only eq. (6). The mutual influence between subspaces, caused by the residual interaction, is contained in the moments of the subspace distributions. This can be easily seen, for example, by examining the variance. For a given subspace α,

$$\sigma_\alpha^2 = \frac{1}{d_\alpha} \sum_{i \in \alpha} \langle i | H^2 | i \rangle = \frac{1}{d_\alpha} \sum_{i \in \alpha} \sum_j \langle i | H | j \rangle \langle j | H | i \rangle. \quad (12)$$

Since H can connect a state $|i\rangle$ inside subspace α to states $|j\rangle$ inside or outside that subspace, the variance can be separated into two parts,

$$\sigma_\alpha^2 = \frac{1}{d} \sum_{i \in \alpha} \left(\sum_{j \in \alpha} \langle i | H | j \rangle \langle j | H | i \rangle + \sum_{j \notin \alpha} \langle i | H | j \rangle \langle j | H | i \rangle \right) \equiv \sigma_{\alpha\alpha}^2 + \sigma_{\alpha\beta}^2. \quad (13)$$

Physically we can interpret the *external width*, $\sigma_{\alpha\beta}$, as the contribution to the spread of the distribution in subspace α due to other subspaces, whereas the *internal width*, $\sigma_{\alpha\alpha}$, represents the mutual influence between states within the subspace. More complicated correlations between different subspaces are given by the higher-order moments. In this way, it is not hard to see that a substantial amount of information concerning the space is present in the low-order moments. Furthermore, one can decompose the moments into even finer parts than we have done in eq. (13) for the variance. By calculating, *e.g.*, the variance due to each external subspace separately, we can examine in detail the mutual interplay between different subspaces.

Another important reason for partitioning the space is that, in nuclear structure studies, we are often interested only in the low-lying part of the spectrum. For large spaces, this means that we are primarily concerned with regions far away from the distribution centroid. In such regions, the function is extremely sensitive to the higher-order moments, *e.g.*, those beyond the third and fourth. Technically, it is usually more difficult to calculate the higher-order moments. By subdividing the space, the complete distribution becomes a sum over many smaller distributions. If the subdivision is taken properly, the centroids of the various subspace distributions can be made to spread over a wide energy region, and a large part of the width of the entire distribution may be taken up by the spread in the centroids of the various subspaces. When this is true, the average variance of the subspace distributions will be smaller than that of the entire distribution, as can be seen from eq. (11). Consequently, each subspace distribution is, on the average, sharper than $\rho(x)$, and the low-lying region is no longer far away from the centroids of the subspaces that are important to it. Once again we need only low-order moments, albeit those of the subspaces.

As far as eigenvalue distributions are concerned, $\rho(x)$ can be found directly in terms of the eigenvalues. For example, we can group the eigenvalues according to energy into bins and plot the number of eigenvalues in each bin in the form of a histogram as, *e.g.*, in Fig. II-1. On the other hand, each eigenstate is, in general, made up of a linear combination of contributions from various subspaces, and as a result the corresponding distribution in a given subspace cannot be obtained from a knowledge of the eigenvalues alone: information from the eigenvectors is also required. If the complete set of eigenvectors $\{\Psi_i\}$ is known, and if each eigenvector is expressed in terms of a linear combination of basis wave functions Φ_j in the form

$$\Psi_i = \sum_{j=1}^{d} c_{ij}\, \Phi_j\,,$$

the strength of the subspace α in the energy range $(E \pm \Delta E/2)$ can be written in the form

$$\rho_\alpha(E)\,\Delta E = \frac{1}{d_\alpha} \sum_{j \in \alpha} \sum_i{}' c_{ij}^2\,, \tag{14}$$

where the prime indicates that the summation over i is restricted to eigenstates in the energy range $(E \pm \Delta E/2)$. However, it is usually extremely tedious, if not impossible, to obtain all the eigenvectors, especially when the active space is large. One of the aims of statistical spectroscopy is to obtain the distributions of a quantity in subspaces directly from the low-order moments without having to solve the complete eigenvalue problem.

Although there is no formal proof, it is easy to see that the density distribution should also be nearly Gaussian if a subspace is made up of a group of basis states with a sufficient number of degrees of freedom. In virtue of this, we have some latitude in choosing the way to partition the space. On the other hand, as we shall see in Chapter IX, some group-theoretical structure in the subdivision is necessary in order to be able to evaluate the moments in the subspaces conveniently. This is important since the main advantage of the statistical approach will be lost if one must spend a large amount of effort to evaluate the necessary moments.

III.3   Distribution of expectation values

In addition to eigenvalues, it is also interesting to study the distribution of the expectation value $\langle E|\hat{O}|E\rangle$ of an operator $\hat{O}$. Besides the familiar electromagnetic moments of a nucleus, sum rule quantities are also examples of expectation values.

The non-energy weighted sum rule quantity $G_0(E)$ is the sum of excitation strengths $R(E', E)$, defined in eq. (II-33), from a given starting state at energy E to all the final states $|E'\rangle$. It can be written in the form

$$G_0(E) = \sum_{E'} R(E', E) = \sum_{E'} \langle E|\hat{O}^\dagger|E'\rangle \langle E'|\hat{O}|E\rangle = \langle E|\hat{O}^\dagger\hat{O}|E\rangle, \quad (15)$$

where we have used the closure relation to obtain the final result. Since we have summed over the final states, $G_0(E)$ depends only on the energy E of the initial state. More generally, we can define the energy weighted sum rule of order p as

$$G_p(E) = \sum_{E'} E'^P R(E', E) = \sum_{E'} \langle E|\hat{O}^\dagger H^P|E'\rangle \langle E'|\hat{O}|E\rangle = \langle E|\hat{O}^\dagger H^P\hat{O}|E\rangle.$$
$$(16)$$

Linear and quadratic energy-weighted sum rules are the most common ones encountered in nuclear physics applications.

For simplicity we shall use the notation

$$K(E) = \langle E|\hat{K}|E\rangle, \quad (17)$$

for the expectation value of an operator $\hat{K}$ as a function of energy. For static moments, for example, $\hat{K}$ is the electromagnetic multipole operator, while for sum rule quantities, $\hat{K} = \hat{O}^\dagger\hat{O}$ for $G_0(E)$, and $\hat{K} = \hat{O}^\dagger H^P\hat{O}$ for $G_p(E)$.

To take advantage of the statistical spectroscopy approach, it is necessary to express eqs. (15) and (16) in terms of traces. For convenience, we shall also make use of average traces, traces divided by the number of states in the space. To distinguish between the two quantities, we shall use $\ll \hat{O} \gg$ for the trace of an operator $\hat{O}$, and

$$\langle \hat{O} \rangle = \frac{1}{d} \ll \hat{O} \gg$$

for the average trace.

The trace of $\delta(H - E)$ is the number of states per unit energy interval at energy E

$$\ll \delta(H - E) \gg \, = I(E) \, , \tag{18}$$

where $I(E)$ is given by eq. (9), and is in general different from unity. The delta function can be expanded in terms of orthogonal polynomials $P_\mu(x)$ in the form

$$\delta(x - y) = \rho(x) \sum_{\mu=0}^{\infty} P_\mu(x) \, P_\mu(y) \, , \tag{19}$$

where the polynomials $P_\mu(x)$ satisfy the relation

$$\int_{-\infty}^{+\infty} P_\mu(x) \, P_\nu(x) \, \rho(x) \, dx = \delta_{\mu\nu} \, . \tag{20}$$

Note that the density distribution $\rho(x)$ is used as the weight function. When the density $\rho(x)$ is Gaussian, we have

$$P_\mu(x) \xrightarrow[\rho \to \rho_G]{} \frac{1}{\sqrt{\mu!}} \, He_\mu(x) \, , \tag{21}$$

as can be seen by comparing eqs. (4) and (20).

A polynomial of order μ is a power series of the argument up to a maximum power μ. If the moments M_ν of $\rho(x)$ are known up to order 2μ, we can find all the polynomials $P_\nu(x)$ up to order μ. Let us first illustrate how the polynomials are obtained by working out explicitly the lowest few orders. Since the $P_\mu(x)$ are normalized according to eq. (20),

$$P_0(x) = 1. \tag{22}$$

Next we can find $P_1(x)$ using the orthonormality condition (20),

$$\int_{-\infty}^{+\infty} P_1(x) \, P_0(x) \, \rho(x) \, dx = 0, \qquad \int_{-\infty}^{+\infty} P_1(x) \, P_1(x) \, \rho(x) \, dx = 1. \tag{23}$$

Since $\rho(x)$ is centered, $\int x \, \rho(x) \, dx = 0$, and we obtain

$$P_1(x) = x. \tag{24}$$

The second-order polynomial has the form

$$P_2(x) = a + bx + cx^2, \tag{25}$$

where a, b, and c are coefficients to be determined by using eq. (20). The orthogonality to $P_0(x)$ yields

$$\int_{-\infty}^{+\infty} P_2(x) \, P_0(x) \, \rho(x) \, dx = a + bM_1 + cM_2 = a + c = 0, \tag{26}$$

and with $P_1(x)$,

$$\int_{-\infty}^{+\infty} P_2(x)\,P_1(x)\,\rho(x)\,dx = aM_1 + bM_2 + cM_3 = b + cM_3 = 0. \quad (27)$$

Eqs. (26) and (27) provide two of the three equations required to determine the three unknown coefficients. The third equation comes from the normalization of $P_2(x)$,

$$\int_{-\infty}^{+\infty} (a + bx + cx^2)^2 \rho(x)\,dx = a^2 + 2abM_1 + (b^2 + 2ac)M_2 + 2bcM_3 + c^2M_4$$

$$= a^2 + (b^2 + 2ac) + 2bc\,M_3 + c^2 M_4 = 1. \quad (28)$$

From eq. (28), we see that moments up to $M_{2\mu}$ are needed to determine $P_\mu(x)$. In general, we can express a polynomial of arbitrary order in the form of a determinant (Cramer, 1946),

$$(D_\mu D_{\mu-1})^{\frac{1}{2}} P_\mu(x) = \begin{vmatrix} 1 & M_1 & M_2 & \cdots & M_\mu \\ M_1 & M_2 & M_3 & \cdots & M_{\mu+1} \\ \vdots & \vdots & \vdots & \ddots & \vdots \\ M_{\mu-1} & M_\mu & M_{\mu+1} & \cdots & M_{2\mu-1} \\ 1 & x & x^2 & \cdots & x^\mu \end{vmatrix}, \quad (29)$$

where D_μ is the same determinant as on the right hand side of the equation except that the last row is replaced by $(M_\mu, M_{\mu+1}, \ldots, M_{2\mu})$.

Using eq. (19) we can now express $K(E)$ in terms of traces. Starting from eq. (17), we have

$$K(E) = \frac{1}{I(E)} \sum_W \langle W | \hat{K}\delta(H - E) | W \rangle = \frac{1}{I(E)} \ll \hat{K}\delta(H - E) \gg . \quad (30)$$

With the help of eq. (19), the delta function is replaced by an orthogonal polynomial expansion,

$$K(E) = \frac{1}{d} \sum_\mu \ll \hat{K} P_\mu(H) \gg P_\mu(E) = \sum_\mu \langle \hat{K} P_\mu(H) \rangle P_\mu(E). \quad (31)$$

In the last step, we have absorbed the dimension d by replacing the trace by an average, and it is understood that both H and E are measured in units of σ and with origin at the distribution centroid.

It is perhaps easier to see the implication of eq. (30) by writing out the first few terms explicitly,

$$K(E) = \langle \hat{K} \rangle + \langle \hat{K}H \rangle E + \langle \hat{K}P_2(H) \rangle P_2(E) + \cdots. \quad (32)$$

The first term is the average of the operator over the entire space. It is the best estimate one can give for the expectation value of $\hat{K}$ for an arbitrary energy E unless one has some further knowledge of the distribution. An improved value can be obtained by adding a linear energy dependence if the correlation of $\hat{K}$ with H is known. If $\hat{K}$ is only weakly correlated with H, we do not expect $K(E)$ to vary appreciably with E. On the other hand, if $\langle \hat{K}H \rangle$ is negative, we expect an increase of $K(E)$ at low energies (below the centroid) over and above $\langle \hat{K} \rangle$ with a corresponding decrease at the higher energy side. Conversely, a positive correlation between $\hat{K}$ and H moves the strength from low to high energy regions. The quadratic energy dependence is contained in the third term in the form of the second-order polynomial $P_2(E)$, and the more complicated energy dependences are provided by the higher-order correlations in the subsequent terms. The use of an orthogonal polynomial expansion normalized with the density distribution as the weight function ensures that the expansion is rapidly convergent.

III.4   Distribution of excitation strengths

The excitation strength, $R(E', E)$, is a function of both the starting state energy E and the final state energy E'. The distribution of $R(E', E)$ is therefore a two-dimensional one in the variables E and E'. Our interest here lies only in the smooth variation of the distribution with the state-to-state fluctuations removed, *e.g.*, by a running or local average.

In addition to the dependence of $R(E', E)$ itself on the energies, the number of states $I(E)$ in the initial space, and $I'(E')$ in the final space, also changes with the energy because of variations in the state densities. Hence the strength function, the total strength measured between two given energy intervals,

$$S(E', E) = I(E)\, I'(E')\, R(E', E), \tag{33}$$

varies with the energies in a way that is in general different from $R(E', E)$.

Given the density distributions, the conversion between $R(E', E)$ and $S(E', E)$ is straightforward. However, in statistical spectroscopy $R(E', E)$ is the quantity that is calculated and $S(E', E)$ is obtained from it via eq. (33). There is occasional confusion between the two quantities since $S(E', E)$ is the quantity usually measured in experiments.

Since it depends on both E and E', the distribution of $R(E', E)$ requires a double orthogonal polynomial expansion, one in E and one in E'. We can take the same approach as for the expectation value by using eq. (19). However, before we can carry out the expansion, we must first express the square of a matrix element as an expectation value, again by the use of a delta function,

$$R(E', E) = \langle E|\hat{O}^\dagger|E'\rangle\langle E'|\hat{O}|E\rangle = \frac{1}{I'(E')} \sum_W \langle E|\hat{O}^\dagger \delta(H - E')|W\rangle \langle W|\hat{O}|E\rangle$$

$$= \frac{1}{I'(E')} \langle E|\hat{O}^\dagger \delta(H - E')\, \hat{O}|E\rangle. \qquad (34)$$

The expectation value can be transformed, in turn, into a trace with the help of a second delta function, and then into a polynomial series in the same way as in eq. (30),

$$R(E', E) = \frac{1}{I(E)\, I'(E')} \ll \hat{O}^\dagger \delta(H - E')\hat{O}\delta(H - E) \gg$$

$$= \frac{1}{dd'} \sum_{\mu\nu} \ll \hat{O}^\dagger P'_\mu(H)\, \hat{O} P_\nu(H) \gg P'_\mu(E')\, P_\nu(E)$$

$$= \frac{1}{d'} \sum_{\mu\nu} \langle \hat{O}^\dagger P'_\mu(H)\, \hat{O} P_\nu(H) \rangle\, P'_\mu(E')\, P_\nu(E), \qquad (35)$$

where $P_\nu(E)$ is the polynomial of order ν defined in the E space, and $P_\mu(E')$ is defined in the E' space.

The first term in eq. (35) is $\langle \hat{O}^\dagger \hat{O} \rangle$, the average strength in the space. The linear energy dependences, described by the next three terms, are given by the correlations of $\hat{O}$ and $\hat{O}^\dagger$ with the Hamiltonian, $\langle \hat{O}^\dagger H \hat{O} \rangle E'$, $\langle \hat{O}^\dagger \hat{O} H \rangle E$, and $\langle \hat{O}^\dagger H \hat{O} H \rangle EE'$. Let us examine one of these coefficients in more detail, for example, $\langle \hat{O}^\dagger H \hat{O} H \rangle$. The average trace is taken over the product of four operators. On the extreme right, we have the Hamiltonian acting in the subspace containing the initial state. The intermediate states generated by the action of this Hamiltonian remain, in general, in the same space, the E space in this case. The effect of this H, therefore, provides the mutual influence between a pair of states in the starting space. The excitation operator $\hat{O}$ to its left takes the system into the final or E' space and the second H supplies the interaction between a pair of states in the final space before $\hat{O}^\dagger$ brings the system back to the starting space. More complicated interplays between the initial and final spaces are described by the higher-order polynomial terms. Again we expect that the action of the first few terms in eq. (35) contains enough mutual influences between the operators and spaces to give an adequate description of $R(E', E)$.

III.5   Partitioning of the space and strength distributions

The importance of the correlation between the Hamiltonian and the excitation operator is even more evident when we partition the space. For simplicity, we shall only discuss the case of configuration subdivision. Each possible arrangement of the m nucleons into the various active shell model orbits is labelled by $m \equiv (m_1, m_2, \ldots, m_\Lambda)$ where m_r is the number of particles in orbit r. In each subspace, we have a centroid,

$$C_m = \langle H \rangle^{(m)} , \tag{36}$$

given by the average trace of H in the configuration, and a density of states $\rho_m(E)$. Analogous to eqs. (9) and (18), we have

$$I_m(E) = d_m \rho_m(E) = \ll \delta(H - E) \gg^{(m)} . \tag{37}$$

Where necessary, we shall use a superscript inside a set of round brackets, as done in eqs. (36) and (37), to indicate the space over which the trace is taken.

The expectation value distribution can be written as a sum over the subspace distributions. Starting from eq. (30), we have

$$K(E) = \frac{1}{I(E)} \ll \hat{K}\delta(H - E) \gg = \frac{1}{I(E)} \sum_m \ll \hat{K}\delta(H - E) \gg^{(m)} , \tag{38}$$

where the trace is taken over all the states within the particular subspace. The delta function can be expanded in terms of orthogonal polynomials $P_\mu^{(m)}(x)$ defined in the subspace m using eq. (29) with the moments of the subspace distribution $\rho_m(E)$ as the input. Similar to eq. (31), we have

$$K(E) = \frac{1}{I(E)} \sum_m \sum_\mu \rho_m(E) \ll \hat{K} P_\mu^{(m)}(H) \gg^{(m)} P_\mu^{(m)}(E)$$

$$= \sum_m \frac{I_m(E)}{I(E)} \sum_\mu \langle \hat{K} P_\mu^{(m)}(H) \rangle^{(m)} P_\mu^{(m)}(E), \tag{39}$$

for the polynomial expansion of the expectation value.

For a zeroth-order theory, we can approximate the summation over μ by the $\mu = 0$ term alone. The strength in each subspace is then a constant, $\langle \hat{K} \rangle^{(m)}$, and the total distribution reduces to

$$K(E) \approx \sum_m \frac{I_m(E)}{I(E)} \langle \hat{K} \rangle^{(m)} .$$

In contrast, the zeroth-order theory without partitioning, given by the first term of eq. (35), yields a constant value $\langle \hat{K} \rangle$ at all energies. Here, since the average values of $\hat{K}$ in each subspace are, in general, different from each other and the contribution of each subspace at a given energy E is weighted by the ratio $I_m(E)/I(E)$, we have some energy dependence even in the zeroth-order theory. In general, we find that the partitioning of the space enables us to achieve the same accuracy with a lower-order polynomial expansion than without one (Draayer, French, and Wong, 1977).

To expand the distribution of $R(E', E)$ we first need to define the average excitation strength from a configuration m to a configuration m',

$$R(m', m) \equiv \frac{1}{d_m d_{m'}} \sum_{\substack{i \in m \\ j \in m'}} |\langle i|\hat{O}|j\rangle|^2 = \frac{1}{d_m d_{m'}} \ll \hat{O}^\dagger (m - m') \hat{O} \gg^{(m)}$$

$$= \frac{1}{d_{m'}} \langle \hat{O}^\dagger (m - m') \hat{O} \rangle^{(m)} . \quad (40)$$

In the last line the division by $d_{m'}$, the dimension of the m' space, is needed since by changing from a trace to an average in the m space, we have only absorbed one of the two factors from the second to the last equality. By $\hat{O}^\dagger (m - m')$ we mean the part of operator $\hat{O}^\dagger$ that changes m to m' (for a one-body operator at most two orbits can change their occupancies, one by $+1$ and the other by -1), and $\hat{O} = \sum_{m,m'} \hat{O}(m - m')$. With $\langle \hat{O}^\dagger (m - m') \hat{O} \rangle^{(m)}$, the average trace of $\hat{O}^\dagger (m - m')\hat{O}$ in the m-space, we can express $R(E', E)$ as a sum over the strengths in each subspace. Starting from the first line of eq. (35), we have

$$R(E', E) = \frac{1}{I(E)\, I(E')} \ll \hat{O}^\dagger\, \delta(H - E')\, \hat{O}\, \delta(H - E) \gg$$

$$= \frac{1}{I(E)\, I(E')} \sum_{mm'} \ll \hat{O}^\dagger (m - m')\, \delta(H - E')\, \hat{O}\, \delta(H - E) \gg^{(m)}$$

$$= \frac{1}{I(E)\, I(E')} \sum_{mm'} \rho_m(E) \rho_{m'}(E')$$

$$\times \sum_{\mu\nu} \ll \hat{O}^\dagger (m - m') P_\mu^{(m')}(H)\, \hat{O} P_\nu^{(m)}(H) \gg^{(m)}\, P_\mu^{(m')}(E') P_\nu^{(m)}(E)$$

$$= \sum_{mm'} \frac{I_m(E) I_{m'}(E')}{I(E)\, I(E')} \frac{1}{d_{m'}}$$

$$\times \sum_{\mu\nu} \langle \hat{O}^\dagger (m - m') P_\mu^{(m')}(H)\, \hat{O} P_\nu^{(m)}(H) \rangle^{(m)}\, P_\mu^{(m')}(E') P_\nu^{(m)}(E). \quad (41)$$

The expression inside the summation over $\mu\nu$ gives the contribution to the strength between E and E' from the subspaces m and m'. Let us represent this by

$$R(m'E', mE) \equiv \frac{1}{d_{m'}} \sum_{\mu\nu} \langle \hat{O}^\dagger (m - m')\, P_\mu^{(m')}(H)\, \hat{O} P_\nu^{(m)}(H) \rangle^{(m)}$$

$$\times P_\mu^{(m')}(E') P_\nu^{(m)}(E). \quad (42)$$

Eq. (41) can then be written in the form

$$R(E', E) = \sum_{mm'} \frac{I_m(E) I_{m'}(E')}{I(E)\, I'(E')} R(m'E', mE), \qquad (41')$$

displaying clearly that the distribution is the sum over the contributions from subspaces weighted by the densities of states in the initial and final spaces.

III.6   Fluctuations and error estimates

The orthogonal polynomial expansions we have just derived for expectation values, eqs. (31) and (39), and excitation strengths, eqs. (35) and (41), are in principle exact if the summations are carried out to infinite order (or to the same order as the dimensions of the spaces involved in finite dimensional cases). However, in practice the necessary input moments are available only up to some low orders. In statistical spectroscopy the basic philosophy is to study the physics using only smooth distributions given by the low-order moments. On the other hand, it is a good idea to have some estimate of the magnitude of the fluctuations involved, even though one may not be able to extract any information from the fluctuations themselves. In the following discussion we shall attempt to give an estimate using the same low-order polynomial expansion as used for the evaluation of the secular variation.

The estimate is based on two assumptions, essentially the same ones as involved in obtaining the Porter-Thomas distribution in the previous chapter. First, the matrix element $x \equiv \langle E'|\hat{O}|E \rangle$ of the excitation operator $\hat{O}$ is taken as a variable with a Gaussian distribution centered around zero,

$$p(x)\, dx = \frac{1}{\sigma\sqrt{2\pi}}\, \exp\left(-\frac{x^2}{2\sigma^2}\right).$$

The variance of the distribution is given by

$$\sigma^2 = \int_{-\infty}^{+\infty} x^2\, p(x)\, dx.$$

We shall also need the fact that, since the distribution is Gaussian, the fourth moment is given by the variance in the form

$$\int_{-\infty}^{+\infty} x^4\, p(x)\, dx = 3\sigma^4 = 3\left\{ \int_{-\infty}^{+\infty} x^2\, p(x)\, dx \right\}^2. \qquad (43)$$

The second assumption is that the excitation strengths are uncorrelated, *i.e.*, the average of the product is equal to the product of the averages of the individual quantities. In particular, we require that the local average of the product of two excitation strengths between two different pairs of energies is equal to the product of the local averages of each of the excitation strengths. Using a bar over a quantity to indicate local averaging as in eq. (II-39), we can express the assumption in the form

$$\overline{R(E_1', E_1)R(E_2', E_2)} = \overline{R(E_1', E_1)}\ \overline{R(E_2', E_2)}. \tag{44}$$

By local averaging we mean an average of the quantity over a small energy domain ΔE,

$$\overline{R(E', E)} = \frac{1}{d_{\Delta E}(E)}\frac{1}{d_{\Delta E'}(E')}\sum_{E_i E_j}{}' R(E_j, E_i)\,,$$

where E_i is restricted in the range $(E \pm \Delta E/2)$ and E_j in $(E' \pm \Delta E'/2)$. Such an averaging procedure takes out the state-to-state fluctuations but preserves the secular variation with energy. In terms of a polynomial expansion, this is equivalent to an expansion only up to some low order.

We shall now make use of the two above assumptions to deduce the variance of the strength distribution at a given energy E. Let us start by evaluating the correlation between moments of orders p and q,

$$
\begin{aligned}
\overline{\mathcal{M}_p(E)\mathcal{M}_q(E)} &\equiv \sum_{W_1, W_2} W_1^p\, W_2^q\, \overline{R(W_1, E)\, R(W_2, E)} \\
&= \left(\sum_{W_1 \neq W_2} + \sum_{W_1 = W_2}\right) W_1^p\, W_2^q\, \overline{R(W_1, E)\, R(W_2, E)} \\
&= \sum_{W_1} W_1^p\, \overline{R(W_1, E)}\, \sum_{W_2} W_2^q\, \overline{R(W_2, E)} \\
&\quad + \sum_{W_1} W_1^{p+q}\left(\overline{R(W_1, E)^2} - \left\{\overline{R(W_1, E)}\right\}^2\right), \tag{45}
\end{aligned}
$$

where in arriving at the last equality we have made use of the independence assumption given by eq. (44) for the first term. In the second term, there are contributions with $W_1 \neq W_2$ in $\overline{R(W_1, E)^2}$ and these are subtracted out explicitly in the last line. We can now define the covariance between the moments of orders p and q as

$$
\begin{aligned}
\Sigma_{p,q}^2(E) &\equiv \overline{\mathcal{M}_p(E)\mathcal{M}_q(E)} - \overline{\mathcal{M}_p(E)}\ \overline{\mathcal{M}_q(E)} \\
&= \sum_{W_1} W_1^{p+q}\left(\overline{R(W_1, E)^2} - \left\{\overline{R(W_1, E)}\right\}^2\right) \\
&\longrightarrow d'\int x^{p+q}\rho(x)\left(\overline{R(x, E)^2} - \left\{\overline{R(x, E)}\right\}^2\right)\,dx \\
&= 2d'\int x^{p+q}\rho(x)\left\{\overline{R(x, E)}\right\}^2\,dx, \tag{46}
\end{aligned}
$$

where we have made use of eq. (43) in arriving at the final result.

It is convenient to carry out the rest of the derivation in terms of polynomial moments defined by

$$\Lambda_\alpha(E) \equiv \sum_W R(W, E) P'_\alpha(W) \longrightarrow d' \int R(x, E)\, \rho(x)\, P'_\alpha(x)\, dx, \qquad (47)$$

where d' is the dimension of the final space. The zeroth-order polynomial moment is simply the integral of $R(x, E)$ over x,

$$\Lambda_0(E) = d' \int R(x, E)\, \rho(x)\, dx. \qquad (48)$$

On substituting the orthogonal polynomial expansion of $R(x, E)$ given by eq. (35), we obtain from the last line of eq. (47),

$$\begin{aligned}
\Lambda_\alpha(E) &= \int \sum_{\mu\nu} \langle \hat{O}^\dagger P'_\mu(H)\, \hat{O} P_\nu(H) \rangle\, P'_\mu(x)\, P_\nu(E)\rho(x)\, P'_\alpha(x)\, dx \\
&= \sum_{\mu\nu} \langle \hat{O}^\dagger P'_\mu(H)\, \hat{O} P_\nu(H) \rangle\, P_\nu(E)\, \delta_{\alpha\mu} \\
&= \sum_\nu \langle \hat{O}^\dagger P'_\alpha(H)\, \hat{O} P_\nu(H) \rangle\, P_\nu(E), \qquad (49)
\end{aligned}$$

where the Kronecker delta in the second line is obtained from the orthogonality condition of the polynomials given by eq. (20). We can now write the excitation strength distribution in the form

$$R(E', E) = \frac{1}{d'} \sum_\mu \Lambda_\mu(E)\, P'_\mu(E'), \qquad (50)$$

by making use of the last member of eq. (49).

The integral in the last member of eq. (46) can now be expressed in terms of polynomial moments. For $p = q = 0$, we have, by making use of eq. (50),

$$\begin{aligned}
\int \rho(x)\{\overline{R(x, E)}\}^2 dx &= \int \rho(x)\, \overline{R(x, E)}\, \frac{1}{d'} \sum_\mu \overline{\Lambda_\mu(E)}\, P'_\mu(x)\, dx \\
&= \frac{1}{d'} \sum_\mu \{\overline{\Lambda_\mu(E)}\}^2, \qquad (51)
\end{aligned}$$

where we have used the definition of polynomial moments given in eq. (47) to obtain the second $\overline{\Lambda_\mu(E)}$.

The mean square fluctuation of the strength at energy E is given by $\Sigma_{00}^2(E)$. In units of the locally averaged integrated strength $\overline{G_0(E)}$, we have

$$\frac{\Sigma_{00}^2(E)}{\{\overline{G_0(E)}\}^2} = \frac{2d' \int \rho(x) \{\overline{R(x,E)}\}^2 \, dx}{\{d' \int \rho(x) \overline{R(x,E)} \, dx\}^2} = \frac{2}{d'} \frac{\sum_\mu \{\overline{\Lambda_\mu(E)}\}^2}{\{\overline{\Lambda_0(E)}\}^2}. \tag{52}$$

The summation in the numerator extends to some maximum order $\mu_{\max}$. If there is indeed a clear separation of the secular variation and the fluctuation, the ratio given by eq. (52) should be independent of the value of $\mu_{\max}$ as long as it is larger than the order required to account for the secular variation in the energy dependence of the strength distribution. In fact, we can use eq. (52) to define a quantity $\tilde{d}(E)$ that gives a measure of the effective number of final states or open channels accessible to a starting state at energy E,

$$\tilde{d}(E) \equiv d' \frac{\{\overline{\Lambda_0(E)}\}^2}{\sum_\mu \{\overline{\Lambda_\mu(E)}\}^2}. \tag{53}$$

Since $\sum_\mu \{\overline{\Lambda_\mu(E)}\}^2 \geq \{\overline{\Lambda_0(E)}\}^2$, we have $\tilde{d}(E) \leq d'$. If all the strength goes into only a small fraction of all the final states, we have a *collective* type of excitation. The effective number of final states is small in this case and we expect the fluctuation to be large since such special features in the strength distribution cannot be expected to persist over a wide range of energy. On the other hand, if the excitation goes to many final states without any special preference to some of the states, the number of open channels is large and the fluctuation is expected to be small.

Higher-order covariances can also be used in principle for such studies. However, very few actual applications are known.

III.7 Convergence and the central limit theorem

So far we have made the assumption that the fluctuation and secular variation are separate phenomena and, as a result, there is a division between the roles of low- and high-order moments. If this is true, a fast convergence for the polynomial expansions of the locally averaged values is expected. There is no known proof that this is true for nuclei although experience and numerical modelling seem to confirm this conjecture. It is also very difficult to find some characteristics in either the Hamiltonian or the excitation operator that may enable one to make the separation between the long- and short-range energy behaviours.

On the other hand, the statistical aspects of the problem may well play the dominant role. Since there is, in general, a sufficient number of degrees of freedom in the distributions of eigenvalues, expectation values, and excitation strengths, we expect the central limit theorem (C.L.T.) to play an important role. There are many different ways to state this powerful theorem in statistics. We shall quote here from p. 214 of Cramer (1946):

> *Whatever be the distributions of the independent variables ξ_ν – subject to certain very general conditions – the sum $\xi = \xi_1 + \cdots + \xi_n$ is asymptotically normal (m, σ), where m and σ are given by*
>
> $$m = m_1 + m_2 + \cdots + m_n, \qquad \sigma^2 = \sigma_1^2 + \sigma_2^2 + \cdots + \sigma_n^2.$$

The most direct application of the central limit theorem in statistical spectroscopy is to eigenvalue distributions. We have essentially made use of it in Chapter II to derive the Gaussian distribution of level density for $N \gg m \gg k$. Under these conditions, the k-body Hamiltonian can change the single-particle states of at most k of the m particles. Since the system is dilute, *i.e.*, only a small fraction of the N available single-particle states are actually occupied, very little restriction is encountered in choosing the k final states for the particles. Hence each term in the Hamiltonian forms essentially an independent random variable. So long as the Hamiltonian is sufficiently general, the number of degrees of freedom is large, the Central Limit Theorem applies, and the resulting eigenvalue distribution is Gaussian.

In general, if a Hamiltonian has ν degrees of freedom, it can be shown that the distribution is actually χ_ν^2, a chi-square distribution with ν degrees of freedom. For example, if $H = \hat{Q} \cdot \hat{Q}$, where $\hat{Q}$ is the quadrupole tensor operator with five components, $Q_{\pm 2}$, $Q_{\pm 1}$, and Q_0, the eigenvalue distribution is $\chi_{\nu = 5}^2$ (French and Draayer, 1979), not quite a Gaussian. However, as ν increases, the deviation from a true Gaussian becomes smaller and smaller.

Excitation operators, in general, are also of low particle ranks. As a result, the Central Limit Theorem holds also in a dilute system and a Gaussian distribution is obtained. However, the two parameters, centroid and width, that define the distribution will be different, in general, from those that give the eigenvalue distribution. Since the origin of our energy scale is taken to be at the center of the spectrum, defined by the centroid of the eigenvalue distribution, the centroid of $K(E)$ is given by $\langle \hat{K} \rangle$. If we ignore this term, the spectrum and strength distribution centroids will coincide. The difference in the widths of the two distributions means that, with respect to the eigenvalue distribution, the strength is shifted in one direction above the centroid and in the other direction below the centroid. The shift is proportional to the distance from the centroid and is given by the ratio of the two widths. If $\sigma_K / \sigma_E < 1$, the distribution of the expectation value is narrower than that of the eigenvalue, and wider if $\sigma_K / \sigma_E > 1$. A schematic diagram showing the differences between the two distributions is given in Fig. III-1.

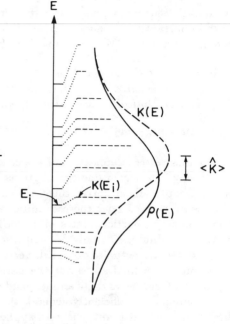

Fig. III-1. Schematic diagram showing the relation between the distributions of eigenvalues $\rho(E)$ and of expectation values $K(E)$.

In terms of the polynomial expansion given by eq. (31), a change in the scale of a distribution is given by the second term $\langle \hat{K}H \rangle E$, as shown explicitly in eq. (32). If the Central Limit Theorem holds, a centroid shift and a scale change are the only differences we can expect between $\rho(E)$ and $K(E)$. Consequently only the first two terms of eq. (32) are needed in describing $K(E)$,

$$K(E) \xrightarrow[\text{C.L.T.}]{} \langle \hat{K} \rangle + \langle \hat{K}H \rangle E. \tag{54}$$

The necessity of including higher-order terms results from the fact that the Central Limit Theorem is not satisfied exactly in a nucleus. On the other hand, so long we are close to the C.L.T. limit, we expect the importance of the higher terms to be insignificant, and fast convergence is then expected.

The same argument applies to the strength distribution $R(E', E)$. Here, analogous to eq. (54), we expect that

$$d'R(E', E) \xrightarrow[\text{C.L.T.}]{} \langle \hat{O}^\dagger \hat{O} \rangle + \langle \hat{O}^\dagger H \hat{O} \rangle E' + \langle \hat{O}^\dagger \hat{O}H \rangle E + \langle \hat{O}^\dagger H \hat{O}H \rangle E'E, \tag{55}$$

and convergence of the sum in eq. (35) can be expected with only the first few terms in realistic cases.

Chapter IV

Simple Averages and Their Propagation

IV.1 Basis representation

Although the trace of an operator is independent of the basis representation used, it is convenient to have a specific set of basis states in mind for the following discussions. Let m be the total number of active particles. The normalized and antisymmetrized m-particle basis wave function, $\Psi(1, 2, \ldots, m)$, can be written as a sum of products of m single-particle wave functions ϕ_i. Where necessary, we assume that the ϕ_i's are the eigenfunctions of a Hamiltonian containing an harmonic oscillator potential and a spin-orbit term. The single-particle states are therefore eigenfunctions of j^2 and j_z, and if isospin is used, t^2 and t_z as well. We shall use the convention that z, the eigenvalue of t_z, is $+1/2$ for a proton and $-1/2$ for a neutron. The eigenvalue of j_z is represented by μ.

The total number N of active single-particle states in the space is assumed to be finite. Where necessary, the single-particle states may be grouped into orbits. In a spherical basis, for example, each orbit contains all the states differing only in the values of μ and z. For identical particles, the number of single particle states in orbit r is therefore $N_r = 2j_r + 1$, and for a mixture of protons and neutrons the number is $N_r = 2(2j_r + 1)$.

IV.2 Dimension of an m-particle space

Before calculating averages, it is instructive to derive first the number of allowed m-particle states in a space defined by N single-particle states. This can be done by a simple, combinatorial counting of the total number of distinct ways to put the m particles into the N available slots. Because of the Pauli exclusion principle, each single-particle state can accommodate only one particle. For the first particle, there are therefore N choices where it may be placed, and for the second one only $(N-1)$ choices are left. By the same argument, the number of choices for particle k is $(N-k+1)$. The total number of possibilities to distribute the m (distinguishable) particles is therefore $N(N-1)(N-2)\cdots(N-m+1)$. However, since the particles are indistinguishable, the order they occupy the various single-particle states is immaterial. As a result, there is an $m!$-fold redundancy in the result above. The correct number of distinct arrangements of m particles in N slots is therefore

$$d_m = \binom{N}{m} = \frac{N!}{m!(N-m)!}, \tag{1}$$

the number of possible m-particle states in the space, or the dimension of the m-particle space.

The m-particle space can be subdivided into a number of subspaces each defined by the occupancy of all the active orbits. Each subspace or *configuration* is labeled by $m \equiv \{m_1, m_2, \ldots, m_A\}$, where m_r is the number of particles in orbit r. The dimension of a configuration is given by the product of the different numbers of ways to arrange m_r particles in N_r single particle states for all the orbits,

$$d_m = \prod_r \binom{N_r}{m_r} \equiv \binom{N}{m}, \tag{2}$$

where we have used the shorthand notation of a binomial coefficient with bold-faced arguments to represent a product of binomial coefficients. Obviously the dimension of the entire space is given by

$$d_m = \sum_m d_m, \tag{3}$$

the sum over all the possible configurations.

It is useful to illustrate the arguments given above by a simple example with only two active orbits. The total number of single-particle states is then $N = N_1 + N_2$. The number of particles in orbit 1 is m_1 and in orbit 2 is m_2. Since m is fixed, $m_2 = m - m_1$; each configuration in this simple example is specified by m_1 alone. The dimension of a configuration is therefore

$$\binom{N_1}{m_1}\binom{N_2}{m_2} = \binom{N_1}{m_1}\binom{N-N_1}{m-m_1}.$$

Eq. (3) implies that

$$\binom{N}{m} = \sum_{m_1} \binom{N-N_1}{m-m_1}\binom{N_1}{m_1}, \tag{4}$$

which is the Vandermonde convolution formula in combinatorial analysis (Riordan, 1968, p. 8). A more general form of the formula can be obtained by replacing N, N_1, m, and m_1 in the above equation by $N-p$, N_1-p, $m-p$, and m_1-p, respectively. Here p is any integer less than or equal to m_1. Since $N - N_1$ and $m - m_1$ are not affected by this replacement we obtain

$$\binom{N - p}{m - p} = \sum_{m_1} \binom{N - N_1}{m - m_1}\binom{N_1 - p}{m_1 - p}, \tag{5}$$

a form which is more convenient for our purpose later.

Returning to the case with an arbitrary number, say Λ, of active orbits, eq. (3) implies a more general convolution relation in the form

$$\binom{N}{m} = \sum_{m} \binom{N - N_1 - N_2 - \cdots - N_{\Lambda-1}}{m - m_1 - m_2 - \cdots - m_{\Lambda-1}}\binom{N_1}{m_1}\binom{N_2}{m_2}\cdots\binom{N_{\Lambda-1}}{m_{\Lambda-1}},$$

a result which can also be derived by applying eq. (5) recursively.

IV.3   One-body operator averages

We can apply the same type of combinatorial arguments as above to find the trace of simple operators. Let us start with $H(1)$, the one-body part of the Hamiltonian. In terms of the single-particle energies ϵ_r, this operator can be written as

$$H(1) = \sum_r \epsilon_r \hat{n}_r, \tag{6}$$

where $\hat{n}_r$ is the number operator for orbit r. The expectation value of $H(1)$ for a state in configuration m is $\sum_r m_r \epsilon_r$. Since all the states in m have the same expectation value for an angular momentum scalar, one-body operator, the trace over m is simply d_m times the expectation value for a state,

$$\ll H(1) \gg^{(m)} = d_m \sum_r m_r \epsilon_r.$$

The trace over the entire m-particle space is obtained by summing over all the configuration contributions

$$\ll H(1) \gg^{(m)} = \sum_m \ll H(1) \gg^{(m)} = \sum_m d_m \sum_r m_r \epsilon_r.$$

To distinguish the two types of trace, we shall call $\ll H(1) \gg^{(m)}$ the scalar trace since it is summed over all the states in the m-particle space, and $\ll H(1) \gg^{(m)}$ the configuration trace since only the states within the configuration m are summed over.

Let us simplify the result of the last equation by dealing first with a two-orbit case. There are two single-particle energies, ϵ_1 and ϵ_2, in this space. The contribution of ϵ_1 to the configuration trace is given by

$$\ll H(1)_{\substack{\epsilon_1 \neq 0 \\ \epsilon_2 = 0}} \gg^{(m)} = d_m m_1 \epsilon_1 = \binom{N_1}{m_1}\binom{N - N_1}{m - m_1} m_1 \epsilon_1,$$

whereas the trace over the entire m particle space is

$$
\begin{aligned}
\ll H(1)_{\substack{\epsilon_1 \neq 0 \\ \epsilon_2 = 0}} \gg^{(m)} &= \sum_{m_1} \binom{N_1}{m_1}\binom{N - N_1}{m - m_1} m_1 \,\epsilon_1 \\
&= N_1 \epsilon_1 \sum_{m_1} \frac{m_1}{N_1} \binom{N_1}{m_1}\binom{N - N_1}{m - m_1} m_1 \\
&= N_1 \epsilon_1 \sum_{m_1} \binom{N_1 - 1}{m_1 - 1}\binom{N - N_1}{m - m_1} m_1 \\
&= N_1 \epsilon_1 \binom{N - 1}{m - 1} = N_1 \epsilon_1 \frac{m}{N}\binom{N}{m} = d_m m \frac{N_1 \epsilon_1}{N}\,.
\end{aligned}
$$

Here we have used eq. (4) to carry out the summation over m_1. The important point of the above equation is that the scalar trace of a one-body operator is simply the product of m and a factor dependent only on the value of the operator in the single-particle space. In terms of an average trace,

$$
\langle H(1)_{\substack{\epsilon_1 \neq 0 \\ \epsilon_2 = 0}} \rangle^{(m)} \equiv \frac{1}{d_m} \ll H(1)_{\substack{\epsilon_1 \neq 0 \\ \epsilon_2 = 0}} \gg^{(m)} = m \frac{N_1 \epsilon_1}{N}\,.
$$

Similarly, we have

$$
\langle H(1)_{\substack{\epsilon_1 = 0 \\ \epsilon_2 \neq 0}} \rangle^{(m)} = m \frac{N_2 \epsilon_2}{N}\,,
$$

for the contribution of ϵ_2 alone.

The result can now be generalized to an arbitrary number of orbits,

$$
\langle H(1) \rangle^{(m)} = m \sum_r \frac{N_r \epsilon_r}{N}\,.
$$

The factor inside the summation is simply the average of $H(1)$ in the $m = 1$ space. In terms of the average single-particle energy $\bar{\epsilon}$ defined by

$$
\bar{\epsilon} = \frac{1}{N} \sum_r N_r \epsilon_r\,, \tag{7}
$$

we have

$$
\langle H(1) \rangle^{(m)} = m \bar{\epsilon}\,. \tag{8}
$$

Physically, we can interpret $\bar{\epsilon}$ as the defining average for $H(1)$ and the factor m in eq. (8) as the *propagator* which connects defining averages with the average in the m-particle space.

Our discussion so far is concerned with the one-body Hamiltonian. It is obvious from the arguments used that the result in eq. (8) applies to any one-body operator whose defining trace is $\bar{\epsilon}$. The result is simple because of the nature of the one-body operator and of the scalar trace. For averages in subspaces, both the defining averages and the propagators will be more complicated, as will be seen in later chapters. However, the basic nature of the propagation procedure for traces and their averages is not different from the simple example illustrated here.

IV.4   Propagation of two-body traces

For a two-body operator the propagation of the scalar average is only slightly more involved. By definition, a two-body operator acts between a pair of particles: hence its trace in m-particle space depends on $\binom{m}{2}$, the number of pairs among m particles, and on $\overline{V}$, the average interaction strength between a pair of particles. The actual expression is given by

$$\langle V \rangle^{(m)} = \binom{m}{2} \overline{V}.$$

This result can be derived in the following way.

Let $|\mu_r\rangle$ and $|\mu_s\rangle$ represent any two of the N single-particle states. A two-body operator V is completely defined if all the two-body matrix elements $\langle \mu_r \mu_s | V | \mu_t \mu_u \rangle$ are known. The trace of V in the two-particle space is given by the sum over all the diagonal matrix elements,

$$\ll V \gg^{(m=2)} = \sum_{\mu_r < \mu_s} \langle \mu_r \mu_s | V | \mu_r \mu_s \rangle,$$

where the restriction $\mu_r < \mu_s$ ensures that the summation is over all different two-particle states in the space. We can therefore define

$$\overline{V} = \frac{1}{d_2} \ll V \gg^{(m=2)} = \binom{N}{2}^{-1} \sum_{\mu_r < \mu_s} \langle \mu_r \mu_s | V | \mu_r \mu_s \rangle, \qquad (9a)$$

as the average trace of the two-body part of the Hamiltonian.

An m-particle state, with the single-particle states $|\mu_1\rangle, |\mu_2\rangle, \ldots, |\mu_m\rangle$ occupied, can be represented by $|\mu_1 \mu_2 \ldots \mu_m\rangle$. By definition, the trace of V in m-particle space is given by

$$\ll V \gg^{(m)} = \sum_{\mu_1 \mu_2 \cdots \mu_m} \langle \mu_1 \mu_2 \ldots \mu_m | V | \mu_1 \mu_2 \ldots \mu_m \rangle,$$

where the sum is restricted to the different m-particle states that can be constructed in the space spanned by the N single-particle states.

Consider first the contribution from the diagonal terms associated with the two-body matrix element $\langle \mu_r \mu_s | V | \mu_r \mu_s \rangle$. Let us define an operator $V_{\mu_r \mu_s}$ which acts only between states μ_r and μ_s by

$$\langle \mu_{r'} \mu_{s'} | V_{\mu_r \mu_s} | \mu_{r''} \mu_{s''} \rangle = \langle \mu_r \mu_s | V_{\mu_r \mu_s} | \mu_r \mu_s \rangle \delta_{rr'} \delta_{ss'} \delta_{rr''} \delta_{ss''}$$

$$\equiv \langle \mu_r \mu_s | V | \mu_r \mu_s \rangle \delta_{rr'} \delta_{ss'} \delta_{rr''} \delta_{ss''}.$$

In the m-particle space,

$$\langle \mu_1 \mu_2 \ldots \mu_m | V_{\mu_r \mu_s} | \mu_1 \mu_2 \ldots \mu_m \rangle = 0 \,,$$

if either μ_r or μ_s is not occupied, and

$$\langle \mu_1 \mu_2 \ldots \mu_m | V_{\mu_r \mu_s} | \mu_1 \mu_2 \ldots \mu_m \rangle = \langle \mu_r \mu_s | V | \mu_r \mu_s \rangle,$$

if both single-particle states μ_r and μ_s are occupied. The total contribution from the two-body matrix element $\langle \mu_r \mu_s | V | \mu_r \mu_s \rangle$ to the m-particle trace is obtained by finding the fraction of all the m-particle states with both μ_r and μ_s occupied.

Since $|\mu_r\rangle$ is one of the N single-particle states in the space and any one of the m particles can occupy it, the probability that a particular m-particle state with $|\mu_r\rangle$ occupied is m/N. Once $|\mu_r\rangle$ is occupied, there are only $(N-1)$ single-particle states and $(m-1)$ particles left. Consequently, the probability of $|\mu_s\rangle$ being occupied in addition to $|\mu_r\rangle$ is $(m-1)/(N-1)$. The joint probability of both $|\mu_r\rangle$ and $|\mu_s\rangle$ being occupied is the product of these two probabilities,

$$\frac{m}{N} \frac{(m-1)}{(N-1)} = \binom{m}{2} \binom{N}{2}^{-1}.$$

The contribution of $\langle \mu_r \mu_s | V | \mu_r \mu_s \rangle$ to the m-particle trace is therefore

$$\ll V_{\mu_r \mu_s} \gg^{(m)} = \binom{N}{m} \binom{m}{2} \binom{N}{2}^{-1} \langle \mu_r \mu_s | V | \mu_r \mu_s \rangle,$$

where the first binomial coefficient comes from the total number of states in the m-particle space. Since only the diagonal part of the operator can contribute to the trace, the general result is given by

$$\ll V \gg^{(m)} = \binom{N}{m} \binom{m}{2} \binom{N}{2}^{-1} \sum_{\mu_r \mu_s} \langle \mu_r \mu_s | V | \mu_r \mu_s \rangle = \binom{N}{m} \binom{m}{2} \overline{V}.$$

In terms of averages instead of traces, we have

$$\langle V \rangle^{(m)} = \binom{m}{2} \overline{V}, \tag{10}$$

the result we wished to demonstrate.

Using eqs. (8) and (10), we obtain the m-particle average for a one-plus two-body Hamiltonian

$$\langle H \rangle^{(m)} = \binom{m}{1} \bar{\epsilon} + \binom{m}{2} \overline{V}, \tag{11}$$

where $\bar{\epsilon}$ and $\overline{V}$ are defined in eqs. (7) and (9) respectively.

It is useful to express $\overline{V}$ in terms of the more familiar two-body matrix elements in the spin-isospin coupled representation. For this purpose we shall regard $|\mu_r\rangle$ to be in orbit r, with angular momentum also represented by r for simplicity, and $|\mu_s\rangle$ in orbit s. The explicit angular momentum rank J of a two-particle state is projected out by the use of Clebsch-Gordan coefficients $\langle r\mu_r s\mu_s | JM \rangle$,

$$|\mu_r \mu_s\rangle = \sum_{JM} \langle r\mu_r s\mu_s | JM \rangle \, |rsJM\rangle, \qquad \mu_r \in r, \ \mu_s \in s.$$

For diagonal matrix elements, we obtain

$$\sum_{\substack{\mu_r \in r \\ \mu_s \in s}} \langle \mu_r \mu_s | V | \mu_r \mu_s \rangle = \sum_{\substack{\mu_r \mu_s \\ JM \\ J'M'}} \langle r\mu_r s\mu_s | JM \rangle \langle r\mu_r s\mu_s | J'M' \rangle \, \langle rsJM | V | rsJ'M' \rangle.$$

Since the Hamiltonian is an angular momentum scalar operator, the matrix element on the left hand side of the above equation vanishes unless $J = J'$ and $M = M'$. As a result,

$$\sum_{\substack{\mu_r \in r \\ \mu_s \in s}} \langle \mu_r \mu_s | V | \mu_r \mu_s \rangle = \sum_{\substack{\mu_r \mu_s \\ JM}} \langle r\mu_r s\mu_s | JM \rangle^2 \, \langle rsJM | V | rsJM \rangle$$

$$= \sum_{JM} \langle rsJM | V | rsJM \rangle,$$

where we have used the orthogonality property of the Clebsch-Gordan coefficients to sum over the projection quantum numbers μ_r and μ_s. Furthermore, since the value of the matrix element is independent of M for an angular momentum scalar operator, we can carry out the summation over M and obtain the result

$$\sum_{\substack{\mu_r \in r \\ \mu_s \in s}} \langle \mu_r \mu_s | V | \mu_r \mu_s \rangle = \sum_J (2J + 1) \, \langle rsJ | V | rsJ \rangle.$$

Using this, we have

$$\overline{V} = \binom{N}{2}^{-1} \sum_{rsJ} (2J + 1) \, \langle rsJ | V | rsJ \rangle, \tag{9b}$$

or

$$\overline{V} = \binom{N}{2}^{-1} \sum_{rsJT} (2J+1)(2T+1) \langle rsJT|V|rsJT \rangle, \qquad (9c)$$

if isospin is included. The statistical factors $(2J+1)$ and $(2T+1)$ appear quite often in subsequent discussions and we shall put

$$[J] \equiv (2J+1), \qquad [JT] \equiv (2J+1)(2T+1),$$

to simplify the notation.

The complete Hamiltonian is the sum of single-particle energy $(k=1)$ and two-body interaction $(k=2)$ parts. In terms of the spherical tensor single-particle creation and annihilation operators defined in Appendix C, it can be written in the form

$$H = \sum_r \epsilon_r \hat{n}_r + \sum_{r \neq s} \epsilon_{rs} [r]^{\frac{1}{2}} \left(A^r \times B^s \right)^0$$

$$- \sum_{\substack{rstu \\ JT}} \varsigma_{rs}\varsigma_{tu} [JT]^{\frac{1}{2}} W_{rstu}^{JT} \left((A^r \times A^s)^{JT} \times (B^t \times B^u)^{JT} \right)^{00}. \quad (12)$$

For simplicity, we shall ignore the contributions from the off-diagonal single-particle energies ϵ_{rs} in the remainder of this chapter. The factors

$$\varsigma_{rr'} = (1 + \delta_{rr'})^{-\frac{1}{2}}$$

are needed so that the defining matrix element for the two-body interaction,

$$W_{rstu}^{JT} \equiv \langle rsJT|V|tuJT \rangle,$$

is the normalized and antisymmetrized two-body matrix element W_{rstu}^{JT} commonly used in the literature. This can be seen from the fact that

$$\langle r's'J'T'|\left((A^r \times A^s)^{JT} \times (B^t \times B^u)^{JT} \right)^{00}|t'u'J'T' \rangle$$

$$= -\frac{\delta_{rr'}\delta_{ss'}\delta_{tt'}\delta_{uu'}\delta_{JJ'}\delta_{TT'}}{\varsigma_{rs}\varsigma_{tu}\sqrt{(2J+1)(2T+1)}},$$

in two-particle space.

IV.5   Propagation for operators of arbitrary particle rank

The arguments used in obtaining the traces of one- and two-body operators can be generalized to a k-body operator $\hat{O}(k)$. An operator whose matrix elements vanish in all $m < k$ particle spaces is a k-body operator or an operator with particle rank k. In addition to particle rank, the operator may also have angular momentum ranks which will be denoted by Greek letters where needed. The operator is completely defined if all the matrix elements in the k-particle space are known. By extending the arguments used above for $k = 1$ (eq. 8) and 2 (eq. 10), we have

$$\langle \hat{O}(k) \rangle^{(m)} = \binom{m}{k} \langle \hat{O}(k) \rangle^{(k)}, \tag{13}$$

where $\langle \hat{O}(k) \rangle^{(k)}$ is the average trace of $\hat{O}(k)$ in the k-particle space.

A mixed particle rank operator $\hat{O}$ of maximum particle rank p can be decomposed in terms of pure rank operators $\hat{O}(k)$,

$$\hat{O} = \sum_{k}^{p} \hat{O}(k).$$

Applying eq. (13) to each term in the sum, we obtain

$$\langle \hat{O} \rangle^{(m)} = \sum_{k}^{p} \binom{m}{k} \langle \hat{O}(k) \rangle^{(k)}. \tag{14}$$

For example, the square of a Hamiltonian is an operator with $k = 2, 3$, and 4. Its average trace is then a sum over $k = 2$ to 4 parts, each one consisting of the product of a propagator and the defining average for the k-body part of H^2.

IV.6   Calculating the defining averages

In order to make use of eq. (14), it is necessary first to decompose a mixed particle-rank operator into its pure rank parts $\hat{O}(k)$. In principle this is an easy operation. In terms of single-particle creation and annihilation operators, a k-body (number-conserving) operator is one with k creation operators on the left of k annihilation operators, i.e., it is in *normal order*. A mixed-particle rank operator $\hat{O}$, with maximum particle rank p, is also made up of p creation and p annihilation operators except that they are not in normal order. By using Wick's theorem, it is possible to rearrange the single-particle operators into normal order form: the commutators generated in the process consist of terms with $(p-1)$, $(p-2)$, ... creation and annihilation operators. In this way, the operator $\hat{O}$ is decomposed into a sum of normal ordered operators, each with a definite particle rank. The procedure, involving the repeated application of commutation relations between single-particle operators, is simple in principle but tedious in practice for any but trivial cases.

However, it is usually not necessary to carry out the decomposition explicitly. This point can be demonstrated by evaluating in the m-particle space the scalar average of an operator $\hat{O}$ with maximum particle rank p by an alternate method from eq. (14). Instead of using $\ll \hat{O}(r) \gg^{(r)}$ as the defining averages, we shall try to express $\ll \hat{O} \gg^{(m)}$ directly in terms of the traces of $\hat{O}$ in r-particle space for $r \leq p$ so as to avoid the decomposition of $\hat{O}$ by particle rank.

In the $m = 0$ space, only $\hat{O}(0)$, the zero-particle rank part of $\hat{O}$, can contribute to the trace. Hence,

$$\langle \hat{O}(0) \rangle^{(0)} = \langle \hat{O} \rangle^{(m=0)} . \tag{15}$$

For $m = 1$, both the $\hat{O}(0)$ and $\hat{O}(1)$ parts can contribute. Using eq. (14), we have

$$\langle \hat{O} \rangle^{(m=1)} = \binom{1}{1} \langle \hat{O}(1) \rangle^{(1)} + \binom{1}{0} \langle \hat{O}(0) \rangle^{(0)}.$$

Substituting the value of $\langle \hat{O}(0) \rangle^{(0)}$ from eq. (15), we obtain

$$\langle \hat{O}(1) \rangle^{(1)} = \binom{1}{1} \langle \hat{O} \rangle^{(1)} - \binom{1}{0} \langle \hat{O} \rangle^{(0)} . \tag{16}$$

Similarly, for $m = 2$, eq. (14) gives

$$\langle \hat{O} \rangle^{(m=2)} = \binom{2}{2} \langle \hat{O}(2) \rangle^{(2)} + \binom{2}{1} \langle \hat{O}(1) \rangle^{(1)} + \binom{2}{0} \langle \hat{O}(0) \rangle^{(0)}.$$

As a result we have

$$\langle \hat{O}(2) \rangle^{(2)} = \binom{2}{2} \langle \hat{O} \rangle^{(2)} - \binom{2}{1} \langle \hat{O} \rangle^{(1)} + \binom{2}{0} \langle \hat{O} \rangle^{(0)}$$

for the defining matrix element of the two-body part of the operator $\hat{O}$.

The process can be continued step by step until we come to some particle number q less than p, the maximum particle rank of the operator $\hat{O}$. We shall now prove by induction that

$$\langle \hat{O}(m) \rangle^{(m)} = \sum_{k=0}^{m} (-1)^{m+k} \binom{m}{k} \langle \hat{O} \rangle^{(k)} \tag{17}$$

is true for all m for $q < m \leq p$. Let us assume that it is true up to $m = q$. Then for $m = q + 1$, we have

$$\langle \hat{O} \rangle^{(m=q+1)} = \sum_{k=0}^{m} \binom{m}{k} \langle \hat{O}(k) \rangle^{(k)}$$

$$= \langle \hat{O}(m) \rangle^{(m)} + \sum_{k=0}^{m-1} \binom{m}{k} \sum_{s=0}^{k} (-1)^{k+s} \binom{k}{s} \langle \hat{O} \rangle^{(k)}$$

$$= \langle \hat{O}(m) \rangle^{(m)} + \sum_{s=0}^{m-1} \sum_{k=s}^{m-1} (-1)^{k+s} \binom{m}{k} \binom{k}{s} \langle \hat{O} \rangle^{(k)} ,$$

where we have started with eq. (14) and use eq. (17) for $\hat{O}(k)$ with $k < m$. The summation over k can be carried out with the aid of the orthogonality relation

$$(-1)^m \delta_{ms} = \sum_{k=s}^{m} (-1)^k \binom{m}{k} \binom{k}{s}$$

$$= (-1)^m \binom{m}{m} \binom{m}{s} + \sum_{k=s}^{m-1} (-1)^k \binom{m}{k} \binom{k}{s},$$

given in eq. (A-13). Since $m > s$ in the last member of the equation above, we have

$$\sum_{r=s}^{m-1} (-1)^r \binom{m}{r} \binom{r}{s} = -(-1)^m \binom{m}{s}.$$

Therefore,

$$\langle \hat{O} \rangle^{(m=q+1)} = \langle \hat{O}(m) \rangle^{(m)} - \sum_{s=0}^{m-1} (-1)^{m+s} \binom{m}{s} \langle \hat{O} \rangle^{(s)},$$

or

$$\langle \hat{O}(m) \rangle^{(m=q+1)} = \langle \hat{O} \rangle^{(m)} + \sum_{s=0}^{m-1} (-1)^{m+s} \binom{m}{s} \langle \hat{O} \rangle^{(s)}$$

$$= \sum_{s=0}^{m} (-1)^{m+s} \binom{m}{s} \langle \hat{O} \rangle^{(s)},$$

and eq. (17) is established by carrying on the process until $m = p$.

For m greater than the maximum particle rank, we start with eq. (14), and on substituting the values of $\langle \hat{O}(s) \rangle^{(o)}$ with the values given by eq. (17), we obtain

$$\langle \hat{O} \rangle^{(m)} = \sum_{k=0}^{p} \binom{m}{k} \langle \hat{O}(s) \rangle^{(s)} = \sum_{k=0}^{p} \binom{m}{k} \sum_{s=0}^{k} (-1)^{k+s} \binom{k}{s} \langle \hat{O} \rangle^{(s)}$$

$$= \sum_{k=0}^{p} \left\{ \sum_{s=k}^{p} (-1)^{k+s} \binom{m}{s} \binom{s}{k} \right\} \langle \hat{O} \rangle^{(k)}. \quad (18)$$

The summation over s inside the curly brackets is carried out in eq. (A-18) of Appendix A. Using that result, we obtain the final form for the scalar average of $\hat{O}$ in m-particle space,

$$\langle \hat{O} \rangle^{(m)} = \sum_{k=0}^{p} \binom{m}{k} \binom{p-m}{p-k} \langle \hat{O} \rangle^{(k)}. \quad (19)$$

The right hand side of this equation involves the averages of the operator $\hat{O}$ itself; as a result, decomposition of $\hat{O}$ by particle rank is no longer necessary. Since we are mostly interested in case with $m > p$, it is more convenient to write eq. (19) in the form

$$\langle\hat{O}\rangle^{(m)} = \sum_{k=0}^{p}(-1)^{p-k}\binom{m-k-1}{p-k}\binom{m}{k}\langle\hat{O}\rangle^{(k)}, \qquad (20)$$

obtained by using eq. (A-5).

The main difference between eqs. (19) and (14) is that two different sets of input traces, $\{\langle\hat{O}\rangle^{(k)}\}$ and $\{\langle\hat{O}(k)\rangle^{(k)}\}$ for $k = 0, 1, 2, \ldots, p$, are used to calculate the m-particle average. They are completely equivalent to each other and either set can be used as the *defining averages* to propagate the m-particle scalar average. The set of input used in eq. (19) is sometimes referred to as the *elementary net* since it is the most straightforward set of input to think of. However, as we see later, it is not necessarily the most convenient set to use.

There are other sets of defining averages which may also be useful. Since the total number of single-particle states is finite, we can decompose $\hat{O}$ in terms of operators $\tilde{O}(\kappa)$ with definite *hole rank*. A κ-hole (number-conserving) operator is one with κ single-particle creation operators to the right of κ single-particle annihilation operators. If $\hat{O}$ has maximum particle rank p, it has also maximum hole rank p. Since the dimension of the single-particle space is N, the space of h holes is the same one as that for $(N - h)$ particles. The defining space of $\tilde{O}(h)$ is therefore that of $(N - h)$ particles. Thus, we can decompose $\hat{O}$ according to hole ranks,

$$\hat{O} = \sum_{h} \tilde{O}(h),$$

and, in analogy with eq. (14), we have

$$\langle\hat{O}\rangle^{(m)} = \sum_{h} \binom{m}{h} \langle\tilde{O}(h)\rangle^{(N-h)}.$$

Similarly, relations equivalent to eqs. (19) and (20) can be written as

$$\langle\hat{O}\rangle^{(m)} = \sum_{r=0}^{p} \binom{m}{r}\binom{p-m}{p-r}\langle\hat{O}\rangle^{(N-r)},$$

and

$$\langle\hat{O}\rangle^{(m)} = \sum_{r=0}^{p}(-1)^{p-r}\binom{m}{r}\binom{m-r-1}{p-r}\langle\hat{O}\rangle^{(N-r)}.$$

As a result, we can use $\{\langle \hat{O} \rangle^{(N-r)}\}$ for $r = 0, 1, 2, \ldots, p$ also as a set of defining averages for the operator $\hat{O}$.

In practical situations, it is more convenient to use a mixture of $\langle \hat{O} \rangle^{(r)}$ and $\langle \hat{O} \rangle^{(N-r)}$ as the defining averages. As long as there are $(p+1)$ independent defining quantities for the scalar averaging of an operator of maximum particle rank p, it is immaterial how they are chosen. One obvious preference is to take the ones that are the easiest to calculate. For most operators, these are likely to be the ones in spaces of the smallest dimensions. The $m = 0$ and $m = N$ spaces have only one state in each and are therefore the simplest cases, and they provide two of the $p+1$ necessary quantities. For $p = 1$, they are adequate for any scalar trace calculations. For higher particle rank operators, we can include the traces in spaces with $m = 2$ and $m = N - 2$, and so on.

We shall now show how to express an m-particle trace in terms of such a mixed set of defining averages. Using eq. (20), we have for $m = N$,

$$\langle \hat{O} \rangle^{(N)} = \binom{N}{p} \langle \hat{O} \rangle^{(p)} + \sum_{r=0}^{p-1} (-1)^{p-r} \binom{N-r-1}{p-r} \binom{N}{r} \langle \hat{O} \rangle^{(r)} .$$

Hence,

$$\langle \hat{O} \rangle^{(p)} = \binom{N}{p}^{-1} \left\{ \langle \hat{O} \rangle^{(N)} - \sum_{r=0}^{p-1} (-1)^{p-r} \binom{N-r-1}{p-r} \binom{N}{r} \langle \hat{O} \rangle^{(r)} \right\} .$$

We can therefore replace $\langle \hat{O} \rangle^{(p)}$ in eq. (20) by the value above and obtain the result

$$\langle \hat{O} \rangle^{(m)} = \binom{m}{p} \binom{N}{p}^{-1} \langle \hat{O} \rangle^{(N)}$$

$$- \sum_{r=0}^{p-1} (-1)^{r-p} \binom{m}{r} \binom{m-r-1}{p-r-1} \frac{N-m}{N-r} \langle \hat{O} \rangle^{(r)} .$$

This equation no longer requires $\langle \hat{O} \rangle^{(p)}$ as input: $\langle \hat{O} \rangle^{(N)}$ is used instead of it. Since $\langle \hat{O} \rangle^{(N)}$ is much easier to obtain than $\langle \hat{O} \rangle^{(p)}$, we have simplified the preparation of the input for calculating the scalar traces of operator $\hat{O}$. Continuing this process, we can use $\langle \hat{O} \rangle^{(N-1)}$ to replace $\langle \hat{O} \rangle^{(p-1)}$ and so on, until we come to $\langle \hat{O} \rangle^{(p-q)}$. The general form is then

$$\langle \hat{O} \rangle^{(m)} = \binom{N-m}{p-q} \sum_{r=0}^{q} (-1)^{r-q} \binom{N-r}{p-q}^{-1} \binom{m-r-1}{q-r} \binom{m}{r} \langle \hat{O} \rangle^{(r)}$$

$$+ \binom{m}{q+1} \sum_{r=0}^{p-q-1} (-1)^{p-q-1-r} \binom{N-r}{q+1}^{-1}$$

$$\times \binom{N-m-r-1}{p-q-r-1} \binom{N-m}{r} \langle \hat{O} \rangle^{(N-r)} . \tag{21}$$

This result was first given by Chang, French, and Thio (1971) and is the fundamental equation for calculating scalar averages in most practical applications. The defining traces of operator $\hat{O}$ now consist of the traces up to particle number q and hole number $p - q - 1$. The choice of q depends somewhat on the nature of the operator itself; in general, $q \approx p/2$.

Chapter V

Diagrammatic Method and Scalar Averaging

In Chapter IV we have illustrated a method of calculating traces by propagation starting with a set of defining quantities of the operator. Although the basic philosophy of the calculation remains unchanged in dealing with more complicated operators, several improvements of the technique are needed before we can evaluate conveniently the traces required in statistical spectroscopy.

In this chapter, we shall be dealing with three different types of operator. At the most fundamental level, we have the creation and annihilation operators for single particles. The operators associated with observables, such as the Hamiltonian operator, the electromagnetic multipole operators, and the nucleon transfer operators, are generally expressed in terms of these single-particle operators: the explicit expressions for some of these operators are given in Appendix D. We shall refer to the operators of this second type as *elementary operators*. As we have seen already in Chapter III, the operators associated with the moments of distributions are generally made up of products of elementary operators. It is for this third type of operator that efficient methods of trace calculation must be devised before statistical spectroscopy can become a practical tool. As far as we are concerned in this chapter, the primary distinction between an elementary and a product operator, is that the former has a set of defining matrix elements associated directly with it, whereas the latter is defined only through its constituent elementary operators.

The diagrammatic method provides a systematic and painless way to obtain expressions for calculating the trace of, in principle, any product operator. However, our interest in statistical spectroscopy is restricted to the low-order moments and we shall confine the discussion to these cases.

V.1 Decomposition of an operator by particle rank

We shall describe in this chapter the diagrammatic method only in terms of scalar averaging where the trace and average trace of an operator is taken over all the states in the m-particle space. The method is, however, more general, and we shall see its extensions to configuration averaging in the next chapter.

The starting point of the diagrammatic method is the commutation relations between the creation operator $a^\dagger$ and the annihilation operator a of a single fermion, which are given by

$$\{a^\dagger, a\} \equiv a^\dagger a + a a^\dagger = 1, \qquad \{a^\dagger, a^\dagger\} = \{a, a\} = 0. \tag{1}$$

If more than one type of single-particle state is involved, a Latin subscript $r, s, \ldots$ will be used to identify the state each operator is associated with. Instead of eq. (1), we now have

$$\{a_r^\dagger, a_s\} = \delta_{rs}, \qquad \{a_r^\dagger, a_s^\dagger\} = \{a_r, a_s\} = 0, \tag{2}$$

as the commutation relations. In the space of zero particles, $|0\rangle$, or particle vacuum, we have $a_s |0\rangle = 0$. As a result, we obtain

$$\langle 0 | a_r^\dagger a_s | 0 \rangle = 0. \tag{3}$$

On the other hand, we have

$$\langle 0 | a_r a_s^\dagger | 0 \rangle = \delta_{rs}, \tag{4}$$

since $a_s^\dagger |0\rangle = |s\rangle$.

We shall now use the commutation relations to carry out the decomposition of an arbitrary operator according to its particle ranks. A number-conserving operator with a definite particle rank p can be written in the form of a normal-ordered product of p single-particle creation operators and p annihilation operators,

$$\hat{O}(p) = V_p \underbrace{a^\dagger a^\dagger \cdots \cdots a^\dagger}_{p} \underbrace{a\, a \cdots \cdots a}_{p}. \tag{5}$$

The subscripts in the single-particle operators have been dropped in the equation to simplify the notation. All operators of particle rank p have the same form: different operators with the same particle rank are distinguished by the value of the defining matrix element V_p. It is obvious from eq. (5) that $\hat{O}(p)$ vanishes in spaces with particle number $m < p$.

An arbitrary operator $\hat{O}$ written in terms of $a^\dagger$ and a can be brought into normal order by moving all the creation operators to the left of all the annihilation operators step by step using the commutation relations given by eqs. (2). For convenience of the discussion, let us use here the symbol α_r to represent either $a_r^\dagger$ or a_r. If two adjacent single-particle operators $\alpha_r \alpha_s$ (anti-)commute, the movement of α_s to the left of α_r produces only an overall negative sign. On the other hand, if they do not (anti-)commute we obtain an additional term produced by the commutator in addition to the sign change. Using eqs. (3) and (4), we can write down the general case as

$$\alpha_r \alpha_s = \, :\!\alpha_r \alpha_s\!: + \, \langle 0 | \alpha_r \alpha_s | 0 \rangle, \tag{6}$$

where the symbol $:\quad:$ stands for the normal-ordered product of the quantity inside. The validity of eq. (6) can be seen by first letting $\alpha_r = a_r^\dagger$ and $\alpha_s = a_s$. Since $\alpha_r \alpha_s$ is already in normal order here, eq. (6) is an identity as the second term on the right hand side vanishes due to eq. (3). Conversely, if $\alpha_r = a_r$ and $\alpha_s = a_s^\dagger$, we have

$$a_r a_s^\dagger = \, :\!a_r a_s^\dagger\!: + \, \langle 0 | a_s a_r^\dagger | 0 \rangle = -a_s^\dagger a_r + \delta_{rs}. \tag{7}$$

This is merely another way of expressing eq. (2).

The second term on the right hand side of eq. (7) comes from the "contraction" of a pair of single-particle operators. We shall refer to this particular type of contraction as a *right-contraction* since the creation operator is originally on the right of the annihilation operator. It is to be distinguished from a *left-contraction* which we shall define later in connection with the *hole* vacuum.

Eq. (7) is a simple application of Wick's theorem. It is extremely useful here as it provides us with a method to extract from $\hat{O}$ the part with a definite particle rank. If p is the maximum particle rank of $\hat{O}$, we have according to the definition of normal ordering,

$$\hat{O}(p) = \, :\!\hat{O}\!:,$$

as the part of $\hat{O}$ of particle rank p. If we apply a single right-contraction on $\hat{O}$ before normal ordering, the result is a $(p-1)$-rank operator. However, the contraction can take place between any pair of single-particle operators in $\hat{O}$ and all the non-vanishing terms contribute to $\hat{O}(p-1)$. Consequently we have

$$\hat{O}(p-1) = \sum_\beta :\!D_R^1(\beta)\hat{O}\!:,$$

where the summation is over all the possible ways to carry out the single right-contraction and one of these, the βth one, is represented by $D_R^1(\beta)$.

In order to obtain the $(p-2)$-rank part, we must apply two right-contractions on $\hat{O}$, and then normal order the result. In general, for $k \leq p$, we have

$$\hat{O}(k) = \sum_\beta :\!D_R^{p-k}(\beta)\hat{O}\!:. \tag{8}$$

In principle, we have achieved the particle rank decomposition required by eq. (IV-14). However, the advantage of eq. (8) goes beyond a method to calculate the trace of the part of $\hat{O}$ with a definite particle rank.

Since the dimension N of the single-particle space is finite, we can simplify the calculation by making use also of the hole vacuum $|N\rangle$, *i.e.*, the space with all the single-particle states occupied. In principle, we could introduce a set of creation and annihilation operators for the holes; however, since their functions are intimately related to those for particle operators, we shall use only the creation and annihilation operators for particles to simplify the notation. By definition, $a_r^\dagger |N\rangle = 0$, since orbit r is already filled. Consequently, we have the analogous results to eqs. (3) and (4)

$$\langle N|a_r^\dagger a_s|N\rangle = \delta_{rs}, \tag{9}$$

$$\langle N|a_r a_s^\dagger|N\rangle = 0. \tag{10}$$

Eq. (9) defines a *left-contraction* with respect to the hole vacuum in the same way as eq. (7) defines a right-contraction with respect to the particle vacuum. The major distinction between a *left-* and a *right-*contraction is that in the former, a pair of single-particle operators with the creation operator on the left is eliminated in the process, while in the latter the pair that is eliminated has the creation operator on the right.

Using $D_L^n(\beta)\hat{O}$ to denote the βth way to contract $\hat{O}$ with n left-contractions, we obtain $\tilde{O}(k)$, the k-hole part of $\hat{O}$, in the form

$$\tilde{O}(k) = \sum_\beta {}^\bullet_\bullet D_L^{p-k}(\beta)\hat{O}{}^\bullet_\bullet. \tag{11}$$

Here we have used the symbol ${}^\bullet_\bullet \quad {}^\bullet_\bullet$ to indicate normal ordering with respect to holes, *i.e.*, particle annihilation operators to the left of all particle creation operators.

In the N-particle space (*i.e.*, the hole vacuum), only the zero-hole part of an operator can contribute to the trace. Hence

$$\langle \hat{O}(k)\rangle^{(N)} = \langle \tilde{O}_k(0)\rangle^{(N)},$$

where $\tilde{O}_k(0)$ is the zero-hole part of $\hat{O}(k)$. However, by eq. (IV-13),

$$\langle \hat{O}(k)\rangle^{(N)} = \binom{N}{k}\langle \hat{O}(k)\rangle^{(k)},$$

and we obtain

$$\langle \tilde{O}_k(0)\rangle^{(N)} = \binom{N}{k}\langle \hat{O}(k)\rangle^{(k)}. \tag{12}$$

Since the dimension of the N-particle space is unity and that of the k-particle space is $\binom{N}{k}$, it is more convenient to carry out the following discussion in terms of traces rather than average traces. Eq. (12) can be rewritten in the form

$$\ll \tilde{O}_k(0) \gg^{(N)} = \ll \hat{O}(k) \gg^{(k)}. \tag{12a}$$

Combining the results of eqs. (8) and (11), we obtain

$$\tilde{O}_k(0) = \sum_\beta {}^\circ_\circ D_L^k(\beta) \sum_\gamma :D_R^{p-k}(\gamma)\hat{O}: {}^\circ_\circ. \tag{13}$$

The final normal ordering with respect to the hole vacuum in eq. (13) is really unnecessary since a zero-hole operator does not contain any single-particle creation and annihilation operators.

The equation is, however, still inconvenient to use since we must normal order the quantity $D_R^{p-k}(\beta)\hat{O}$ before we can apply the k left-contractions. Using eqs. (12a) and (13), we can write

$$\ll \hat{O}(k) \gg^{(k)} = \sum_{\beta\gamma} D_L^k(\beta):D_R^{p-k}(\gamma)\hat{O}:. \tag{14}$$

The right hand side of eq. (14) is now a fully contracted operator since there are a total of p contractions, the same number as the maximum particle rank of $\hat{O}$. As a result, there is no need to take the trace of the right hand side of eq. (14) since there is no longer any single-particle creation and annihilation operator left. The only remaining complication in the expression is the normal ordering after the $(p - k)$ right-contractions. We can avoid this as well if the k left-contractions can be carried out before the normal ordering. Once this is done, normal ordering is no longer necessary as the quantity within : : is fully contracted.

Consider now a particular left-contraction to be applied to a pair of single-particle operator $\alpha_r \alpha_s$ after they are put in normal order. If $\alpha_r = a_r^\dagger$ and $\alpha_s = a_s$, they are in normal order to start with, normal ordering therefore has no effect, and it does not matter whether the left-contraction is performed before or after the normal ordering. On the other hand, if $\alpha_r = a_r$ and $\alpha_s = a_s^\dagger$, the normal ordering will interchange their relative positions. If we wish to contract them before normal ordering, it is necessary to change the left-contraction, which was to be carried out after the normal ordering, to a right-contraction. In addition there is also an *extra* minus sign introduced since it is no longer necessary to interchange the positions of $\alpha_r \alpha_s$ in the normal ordering process as they have been contracted away already.

In general, it is not easy to keep account of the particular pairs of single-particle operators to be contracted. However, since we are summing over all possible k left-contractions in a normal-ordered operator anyway, we have for an arbitrary operator $\hat{O}$ the relation,

$$\sum_\beta D_L^k(\beta):\hat{O}: = \sum_{t=0}^{k}(-1)^{k-t}\sum_{\beta\gamma} :D_L^t(\beta)D_R^{k-t}(\gamma)\hat{O}:. \tag{15}$$

The sum over t takes care of the various possible arrangements between the contracting single-particle operators. Eq. (14) now becomes

$$\ll \hat{O}(k) \gg^{(k)} = \sum_{t=0}^{k} \sum_{\beta\gamma\delta} (-1)^{k-t} D_L^t(\beta) D_R^{k-t}(\gamma) D_R^{p-k}(\delta) \hat{O}, \qquad (16)$$

a form which eliminates the need of normal ordering.

The two groups of right-contractions on the right hand side of eq. (16) can be combined into a single one. There are altogether $(p - t)$ right-contractions in the product $D_R^{k-t}(\eta) D_R^{p-k}(\varsigma)$. However, since they are separated into two groups, an overall factor is needed when they are put into a single group. This factor arises from the following considerations. There are $(p-t)!$ different ways to order the $(p-t)$ contractions. In $D_R^{p-t}(\gamma)$ this ordering is immaterial. On the other hand, in the product $D_R^{k-t}(\eta) D_R^{p-k}(\varsigma)$ the ordering is unimportant only among the $(k-t)!$ possible arrangements in η and the $(p-k)!$ in ς. Hence, we have the relation

$$\sum_{\beta\gamma} D_R^{k-t}(\beta) D_R^{p-k}(\gamma) \hat{O} = \binom{p-t}{k-t} \sum_{\delta} D_R^{p-t}(\delta) \hat{O}.$$

The final form of eq. (14) is then

$$\ll \hat{O}(k) \gg^{(k)} = \sum_{t=0}^{k} (-1)^{k-t} \binom{p-t}{p-k} \sum_{\beta\gamma} D_L^t(\beta) D_R^{p-t}(\gamma) \hat{O}, \qquad (17)$$

which we shall simplify further.

It is convenient to introduce a special symbol, O_t^p, for the fully contracted quantity on the right hand side of eq. (17),

$$O_t^p \equiv \sum_{\beta\gamma} D_L^t(\beta) D_R^{p-t}(\gamma) \hat{O}. \qquad (18)$$

We shall demonstrate that O_t^p can be used as the defining quantity for the product operator $\hat{O}$ in trace calculations in the same way as $\langle \hat{O}(k) \rangle^{(k)}$ and $\langle \hat{O} \rangle^{(k)}$ were used in Chapter IV.

Starting with eq. (IV-13), we have, in terms of traces instead of averages,

$$\ll \hat{O}(k) \gg^{(m)} = \binom{N}{m} \langle \hat{O}(k) \rangle^{(m)} = \binom{N}{m}\binom{m}{k} \langle \hat{O}(k) \rangle^{(k)}$$

$$= \binom{N}{k}^{-1} \binom{N}{m}\binom{m}{k} \ll \hat{O}(k) \gg^{(k)},$$

or

$$\ll \hat{O}(k) \gg^{(m)} = \binom{N-k}{m-k} \ll \hat{O}(k) \gg^{(k)}, \qquad (19)$$

where the combinatorial identity,

$$\binom{N-k}{m-k} = \binom{N}{k}^{-1} \binom{N}{m} \binom{m}{k},$$

can be shown to be true by expanding the binomial coefficients explicitly in terms of factorials. Using eqs. (17), (19), and summing over the particle ranks of $\hat{O}$, we obtain

$$\ll \hat{O} \gg^{(m)} = \sum_{k} \binom{N-k}{m-k} \sum_{t=0}^{k} (-1)^{k-t} \binom{p-t}{p-k} O_t^p$$

$$= \sum_{t=0}^{p} \left\{ \sum_{k=t}^{p} (-1)^{k-t} \binom{N-k}{m-k} \binom{p-t}{p-k} \right\} O_t^p$$

$$= \sum_{t=0}^{p} \binom{N-p}{m-t} O_t^p, \qquad (20)$$

where in arriving at the final result we have used an alternate form of the Vandermonde convolution formula given in eq. (A-17). A check on the equation is provided by calculating the total number of states in the space. In this case, $\hat{O}$ is simply the unit operator; hence $t = p = 0$, $O_t^p = 1$, and we recover the result of eq. (IV-1).

Eq. (20) is equivalent to eqs. (IV-14) and (IV-20) except that O_t^p is used instead as the defining quantity for $\hat{O}$. The quantity O_t^p depends on the nature of the operator $\hat{O}$ and the way it is contracted. If $\hat{O}$ is made up of a product of *elementary* operators, each of which is given by a set of defining matrix elements, then O_t^p can be expressed directly in terms of these defining matrix elements.

A scalar trace calculation is now separated into two parts, the fully contracted quantity O_t^p and the propagator $\binom{N-p}{m-t}$, with the former depending only on the structure of the operator and the particular way it is contracted, and the latter being a simple binomial factor depending on the space over which the trace is taken and the values of p and t.

V.2   Diagrammatic method

For an operator of maximum particle rank p, there are $p!$ different possible ways to contract all the single-particle creation and annihilation operators. This is a number far in excess of the number of independent pieces of information required to specify the input for a trace calculation. A large degree of redundancy must therefore exist in the set $\{O_t^p\}$, and we shall now describe a method to reduce the redundancy to a minimum.

Because of the symmetries in the product operator $\hat{O}$, different ways of contracting the single-particle operators contained in it may be related in a trivial way. It is advantageous to use a diagrammatic method to explore the relationships. For our purpose here, it is most convenient to adopt a scheme based on the Hugenholtz (1957) diagrams so that we can take advantage of the antisymmetrization relation between the single-particle creation and annihilation operators (Chang and Wong, 1978). As we shall see below, the connection with the Goldstone diagrams commonly used in calculations involving perturbation expansions is obvious in most cases.

It is convenient to have a pictorial representation of a diagram. We shall use the scheme that an elementary operator is represented by a • and each of the single-particle operators making up the elementary operator is represented by an arrow emerging from the dot for $a^\dagger$, as shown in Fig. V-1a, and by an arrow going into the dot for a, as shown in Fig. V-1b. A one-body operator of the form $a^\dagger a$ is given in Fig. V-1c with one arrow going in and one coming out. A two-body operator, with two incoming and two outgoing arrows, is given by Fig. V-1d.

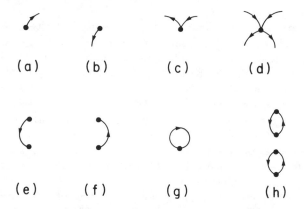

Fig. V-1. Diagrammatic representation of operators. Figure (a) represents a single-particle creation operator $a^\dagger$, (b) a single-particle annihilation operator a, (c) a one-body operator, and (d) a two-body operator. Figure (e) is a left-contraction and (f) a right-contraction. Figures (g) and (h) are examples of a self-contraction and a disconnected diagram, respectively.

A product operator $\hat{O}$ made up of q elementary operators is represented by q dots arranged vertically with the leftmost elementary operator represented by the topmost dot in the diagram. A contraction between a creation operator in elementary operator i with an annihilation operator in elementary operator j is represented by an arrow emerging from dot i and going into dot j. A left-contraction is then indicated by an arrow going downward in the diagram (*i.e.*, elementary operator i is on the left of j) and a right-contraction by an arrow going upward. These are shown in Figs. V-1e and f. A fully contracted operator therefore has all the lines starting and ending at elementary operators.

There are two types of diagram that should be pointed out here. The first type is the self-contraction diagram, given in Fig. 1g, in which both the creation and annihilation operators of the contraction come from the same elementary operator. In this case the direction of the arrow is redundant as it is neither a left-contraction nor a right-contraction. (We shall however represent a self-contraction by D_L to avoid defining a new symbol.) A self-contraction is similar to a number operator and can be treated in a simpler way than other types of contractions. We shall return to this point later in the discussion of unitary decompositions.

A second type is the disconnected diagram. If the contraction pattern is such that the elementary operators are separated into two or more groups without any contractions joining one group to another, the diagram is called disconnected. An example is given in Fig. V-1h. Since the operators are fully contracted within each group, the intermediate state between any two such groups must be identical to the starting state. This is obvious for the simple case shown in Fig. V-1h; other cases can usually be transformed into an equivalent form. Disconnected diagrams therefore represent terms which can be evaluated as the product of two or more terms each simpler than the product. Although we shall not make explicit use of this simplifying feature here, the potentials of possible applications are quite numerous.

V.3   Symmetries and basic diagrams

One of the primary motivations for having a diagrammatic representation of the various terms in O_t^p is to find a convenient way to identify relations between different ways, or *diagrams*, to contract $\hat{O}$. A large amount of effort in evaluating the trace of $\hat{O}$ may be saved if we can identify terms in O_t^p that are either identical in their contributions or similar in structure so they can be evaluated together.

For two diagrams to be related to each other, there must be some symmetry, either in the structure of the product operator or in the nature of the elementary operators. Several different types of symmetry are possible; the particular ones we choose to make use of here are the symmetry in the defining matrix elements of the elementary operators, permutation symmetry between equivalent elementary operators, unitary symmetry, and particle-hole symmetry. In general, each additional symmetry relation reduces the number of independent terms in O_t^p, but the complexity in taking advantage of the symmetry increases. A balance must therefore be established between the complications introduced by one more symmetry and the savings that can be gained from using it. The choice depends to a large extent on the type of average one wishes to calculate.

If two diagrams can be transformed into each other using a set of given symmetries, they are said to be topologically identical. For each set of symmetries, there is a minimum set of distinct diagrams that cannot be transformed into each other. This is the set of *basic diagrams*. The success of a diagrammatic method depends crucially on the ease with which the complete set of basic diagrams involved can be identified.

It is usually quite simple to check whether a particular diagram is a basic one. By definition, a basic diagram cannot be generated from any other basic diagrams using all the available symmetry transformations. From an operational point of view, one can always check whether a new diagram is a basic one by applying all the symmetry operations to the diagram and seeing whether it can be transformed into any of the basic diagrams already known. It is, however, quite a different matter to ensure that a complete set of basic diagrams has been found. To check this, one must verify whether all the $p!$ possible diagrams can indeed be generated from the set. For large values of p, this task can be quite tedious. It is therefore highly advantageous that the diagrammatic representation used can be adapted easily to computer processing. In Appendix E, an algorithm is given by which all the work involved in constructing a complete set of basic diagrams can be carried out efficiently by a machine.

V.4 Symmetries in the defining matrix elements

The defining matrix elements of an elementary operator have symmetries arising from the nature of the operator itself. For example, the (anti-symmetrized) two-body matrix elements W_{rstu}^{JT} of eq. (IV-12) defining the two-body interaction part of a Hamiltonian, have the following symmetries,

$$W_{rstu}^{JT} = -(-1)^{r+s-J-T}W_{srtu}^{JT}$$
$$= -(-1)^{t+u-J-T}W_{rsut}^{JT} = (-1)^{r+s-t-u}W_{srut}^{JT}, \qquad (21)$$

where we have used the orbital indices r, s, *etc.* in the phase factor to stand for both the orbital spin and isospin of the single-particle orbits they are associated with, *i.e.*,

$$(-1)^r \equiv (-1)^{j_r + \frac{1}{2}}.$$

Because of eq. (21), two diagrams differing only in the interchange of indices r and s (and t and u) are simply related to each other by a phase factor. As a result, we need to calculate only one of them and can obtain the other one using the symmetry property. By the same token, there is no need to distinguish between the two single-particle creation operators, and between the two annihilation operators, of a two-body operator in a diagram, such as the one shown in Fig. V-1d. An example is given in Section 8 to provide some of the details in taking advantage of the symmetry of the two-body matrix elements.

Among the elementary operators, the Hamiltonian is encountered most frequently and has more complicated structures. As a result, the most time-consuming part of a calculation usually involves the two-body part of the Hamiltonian. It is therefore useful trying to reduce as much as possible calculations involving the two-body matrix elements. There are also other matrix element symmetries that are potentially useful but seldom invoked in practice. For example, a (one-body) electromagnetic operator, which is usually defined in terms of single-particle values v_{rs}^{λ} as done in Appendix D, has the symmetry

$$v_{rs}^{\lambda} = (-1)^{r-s} v_{sr}^{\lambda},$$

which can be used in conjunction with particle-hole symmetry.

V.5   Permutation symmetry

The number of different types of elementary operator normally encountered in nuclear spectroscopy is small. For example, the ones used in the examples of Chapter X can be classified into the following categories:

[1]	one-nucleon creation operator	$a^{\dagger}$,
[2]	one-nucleon annihilation operator	a,
[3]	two-nucleon creation operator	$a^{\dagger}a^{\dagger}$,
[4]	two-nucleon annihilation operator	aa,
[5]	one-body operator	$a^{\dagger}a$,
[6]	two-body operator	$a^{\dagger}a^{\dagger}aa$.

In addition to these, the three- and four-nucleon creation and annihilation operators ($a^{\dagger}a^{\dagger}a^{\dagger}$, aaa, $a^{\dagger}a^{\dagger}a^{\dagger}a^{\dagger}$, and $aaaa$) are also of physical interest, but we shall not deal with them here.

As for product operators, the general forms usually appearing in statistical spectroscopy consist of an excitation operator $\hat{O}$, together with its adjoint $\hat{O}^{\dagger}$, and several Hamiltonian operators, as shown in Chapter III. A typical one can be written as $\hat{O}^{\dagger}H^r\hat{O}H^s$, as in eq. (III-35). In a trace calculation, terms corresponding only to a different order of the H's among the first group of r and among the second group of s factors, must give the same contribution if all the Hamiltonian operators are identical. Furthermore, permutations of the H's between the two groups cannot produce any difference either although the contraction pattern corresponding to these terms may differ. The diagrams representing these terms must therefore be topologically identical. Although in the present example permutations can take place only between the Hamiltonians, the same argument applies to any other elementary operators that appear more than once in a product operator.

The usefulness of the permutation symmetry is not restricted to identical operators. In fact, even if some of the H's are different from each other, for example, due to the fact they operate in different spaces, the expressions to calculate their contributions to the trace have the same form for all the permutations of the same diagram; only the defining matrix elements entering into the calculations are different. This leads to the concept of *equivalent* operators, *i.e.*, elementary operators that are identical in form but may be different in their defining matrix elements. For all the permutations between equivalent operators that lead to the same basic diagram, the same expression can be used to evaluate their contributions. Furthermore, only one among all the permutations of identical elementary operators needs to be actually calculated since the others contribute the same amount.

V.6   Unitary symmetry

An operator with a fixed particle rank can be decomposed according to its properties under a unitary transformation among the N single-particle basis states. Let us start by considering the simple case of a number-conserving operator of particle rank one.

In eq. (IV-6) the one-body part of the Hamiltonian, $H(1)$, is given in the form

$$H(1) = \sum_r \epsilon_r \hat{n}_r ,$$

where ϵ_r is the single-particle energy and $\hat{n}_r$ is the number operator for orbit r. The number operator $\hat{n}$ in scalar averaging, on the other hand, measures the number of particles $m = \sum_r m_r$ in a state regardless of the distribution of the particles over the different orbits. The expectation value of $H(1)$ in a state with occupancy $m \equiv \{m_1, m_2, \ldots\}$ is given by

$$\langle m|H(1)|m\rangle = \sum_r \epsilon_r m_r = m\bar{\epsilon} + \sum_r (\epsilon_r - \bar{\epsilon})m_r , \qquad (22)$$

where in the last member we have taken out the contribution from the average single-particle energy $\bar{\epsilon}$ given by eq. (IV-7). Hence $H(1)$ can also be written in the form

$$H(1) = \bar{\epsilon}\hat{n} + \sum_r \tilde{\epsilon}_r \hat{n}_r , \qquad (23)$$

where $\tilde{\epsilon}_r \equiv \epsilon_r - \bar{\epsilon}$ is the traceless single-particle energy for orbit r.

A number operator is a one-body operator. However, since it is diagonal in a state with a definite particle number, it is a fully contracted operator with a (self)-contraction between the creation and annihilation operators. Diagrammatically it can be represented by the diagram in Fig. V-1g.

In eq. (23), $H(1)$ is expressed as a sum of two parts. The contribution of the first part depends only on the total number of particles in the state and is therefore invariant under a rotation of the single-particle basis. The remainder of $H(1)$, which cannot be written in terms of a self-contraction, forms the second part. It is a function of the specific number of particles in each orbit and is therefore dependent on the particular single-particle representation used. Under a unitary transformation of the N single-particle states, the first part of $H(1)$ behaves like a scalar and the second part as a rank-one tensor in the space. Similar to tensors in other spaces, an operator with a definite rank has special properties that can be exploited to simplify a calculation. We shall now show how to make use of the properties of unitary tensors in handling traces.

Fig. V-2. Two-body operator with two (a), one (b), and no self-contraction (c).

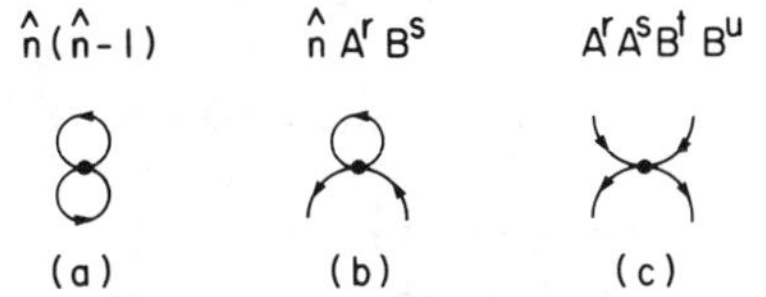

$$\hat{n}(\hat{n}-1) \qquad \hat{n}\, A^r\, B^s \qquad A^r A^s B^\dagger\, B^u$$

(a)   (b)   (c)

A two-body (number-conserving) operator has the form $a^\dagger a^\dagger a a$. It can be decomposed into three parts: one with two self-contractions, one with one self-contraction, and one with no self-contraction, as shown in Fig. V-2. The part with two self-contractions is again a scalar under unitary transformations. However, it cannot be simply a product of two number operators of the form $\hat{n}\cdot\hat{n}$, since a two-body operator, by definition, vanishes in a space with less than two particles, whereas $\hat{n}\cdot\hat{n}$ is a one- plus two-body operator since it does not vanish in the space of one particle. The unitary scalar two-body operator we need must therefore take the form $\hat{n}(\hat{n}-1)$ or $\binom{\hat{n}}{2}$. Similarly, a two-body operator with one self-contraction has the form $(\hat{n}-1)A^r B^s$. For angular-momentum scalar, parity conserving operators, such as $H(2)$, r and s must be radially degenerate orbits for $r \neq s$, i.e., orbits differing only in the principal quantum number.

It is worthwhile to point out here that the form of an operator with a definite unitary rank depends also on its particle rank. If we restrict the discussion to number-conserving operators, as we have done so far in most cases, only a single index is needed to specify the particle rank since the number of single-particle creation operators, N_A, is always equal to the number of annihilation operators, N_B. If these two numbers are different, an additional index,

$$\mu \equiv (N_A - N_B)/2, \tag{24}$$

is required to specify the particle rank unambiguously. When μ is not zero, k is given by

$$k \equiv (N_A + N_B)/2, \tag{25}$$

so that for $N_A = N_B$ the meaning of k remains unchanged.

A general fixed particle-rank operator is therefore labelled by two indices (k, μ). Since a unitary decomposition cannot change μ, only one additional index, ν, is required to indicate the unitary rank. We can now write the unitary decomposition of the particle-rank (k, μ) part of an operator $\hat{O}$ in the form

$$\hat{O}(k, \mu) = \sum_{\nu=|\mu|}^{(k, N-k)_<} \hat{O}_k^\nu(k, \mu), \tag{26}$$

where, on the right-hand side, the superscript ν on $\hat{O}$ indicates the unitary rank and the subscript k the particle rank of the original operator. The operator $\hat{O}_k^\nu(k, \mu)$ has particle rank (k, μ) and is made up of $(k + \mu)$ single-particle creation operators and $(k - \mu)$ annihilation operators with $(k - \nu)$ (self)-contractions between them. It is therefore connected to a chain of unitary operators made up of the product of $\hat{O}_k^\nu(r, \mu)$ times a $(k - r)$-order polynomial in the number operator of the form

$$\hat{O}_k^\nu(k, \mu) = \binom{\hat{n} - \nu - \mu}{k - \nu} \hat{O}_k^\nu(\nu, \mu), \tag{27}$$

used earlier for the simple case of the $(k, \mu) = (2, 0)$ part of the Hamiltonian. It is easy to see that $\hat{O}_k^\nu(\nu, \mu)$, the minimum particle rank member of the chain, is a fully contracted operator in the sense that

$$D_L \hat{O}_k^\nu(\nu, \mu) = 0, \tag{28}$$

since, by definition, if any contraction on $\hat{O}_k^\nu(\nu, \mu)$ does not yield a vanishing result, the operator can be reduced to one of lower unitary rank. Under a rotation of the single-particle basis, $\hat{O}_k^\nu(\nu, \mu)$ belongs to the irreducible representation of $U(N)$ given by a Young tableau of two columns with lengths $(N - \nu + \mu)$ and $(\nu + \mu)$, as described in detail by Chang, French and Thio (1971).

Eq. (28) can be used to carry out the unitary decomposition of fixed-particle rank operators. We shall assume here that the operator $\hat{O}$ has already been decomposed according to particle rank, say, by the use of eq. (8), and the particular part we wish to decompose further according to unitary rank is $\hat{O}(k,\mu)$. The minimum unitary rank ν_m for this operator is then $|\mu|$ (since $\nu \geq \mu$ by definition) and we can extract out this part of $\hat{O}(k,\mu)$ by applying $(k - \nu_m)$ contractions,

$$\hat{O}_k^{\nu_m}(k,\mu) = D_L^{k-\nu_m}\hat{O}(k,\mu). \tag{29}$$

The operator remaining after removing the minimum unitary rank part,

$$\hat{O}'(k,\mu) \equiv \hat{O}(k,\mu) - \hat{O}_k^{\nu_m}(k,\mu), \tag{30}$$

has minimum rank $\nu_m = |\nu|+1$. We can separate out this part of $\hat{O}_k(k,\mu)$ by applying $k - |\nu|+1$ self-contractions to $\hat{O}'(k,\mu)$ in the same way as we have obtained $\hat{O}_k^{\nu_m}(k,\mu)$ from $\hat{O}(k,\mu)$. In this way, all the pure unitary rank parts can be projected out step by step till only the maximum rank $\nu = k$ part is left.

A specific example is perhaps helpful here. The two-body interaction part of the Hamiltonian has been defined in eq. (IV-12). Since it is a $(k,\mu) = (2,0)$ operator, the possible unitary ranks are $(\nu,\mu) = (0,0)$, $(1,0)$, and $(2,0)$. The $(0,0)$ part has the form,

$$\hat{O}_2^0(0,0) = \overline{W}\,\frac{\hat{n}(\hat{n}-1)}{2},$$

according to eq. (27). The question now is to express $\overline{W}$ in terms of the two-body matrix elements W_{rstu}^{JT}. Instead of $a_r^\dagger$ and a_r, it is more convenient to use here the spherical tensor operators A^r and B^r given in Appendix C. The number operator takes the form

$$\hat{n} = \sum_r [r]^{\frac{1}{2}}(A^r \times B^r)^0,$$

where the superscript 0 indicates that the operator is coupled to angular momentum zero. Hence

$$\hat{O}_2^0(0,0) = \frac{\overline{W}}{2}\sum_{rs}\left\{[rs]^{\frac{1}{2}}\left((A^r \times B^r)^0 \times (A^s \times B^s)^0\right)^0 - [r]^{\frac{1}{2}}(A^r \times B^r)^0\delta_{rs}\right\}$$

$$= -\overline{W}\sum_{r\leq s;\,\Gamma}[\Gamma]^{\frac{1}{2}}\,\varsigma_{rs}^{-2}\left((A^r \times A^s)^\Gamma \times (B^r \times B^s)^\Gamma\right)^0, \tag{31}$$

where the last line is obtained by recoupling the single-particle operators using the relations given in Appendix C. Comparing this result with eq. (IV-12), it is obvious that $\overline{W}$ can only involve diagonal two-body matrix elements W_{rsrs}^{JT}, and the weight of each one of them is exactly that for $\overline{V}$ given in eq. (IV-9) for the trace of the two-body part of H. That the two quantities are the same is not accidental since only the unitary scalar part can contribute to the trace of an operator.

The $(\nu, \mu) = (1,0)$ part of $\hat{O}(2,0)$ has the form

$$\hat{O}_2^0(1,0) = \sum_{rs} \overline{W}_{rs}(\hat{n} - 1)[r]^{\frac{1}{2}}(A^r \times B^s)^0. \tag{32}$$

Explicit calculation yields the result

$$\overline{W}_{rs} = \frac{1}{N_r(N-2)} \sum_t \sum_{JT} [\Gamma] W_{rtst}^{\Gamma} \varsigma_{rt}^{-1} \varsigma_{st}^{-1} - \overline{W}, \tag{33}$$

where N_r is the number of single-particle states in orbit r. The $(\nu, \mu) = (2,0)$ part is obtained by subtracting from $\hat{O}(2,0)$ the parts already included in the $(0,0)$ and $(1,0)$ parts given above.

The main advantage of unitary decomposition is that all the diagrams containing one or more self-contractions are reduced in complexity from their corresponding forms without the self-contractions. This comes from the fact that number operators can easily be taken out of the trace calculation and replaced by the particle number: the remainder of the diagram is therefore reduced in particle rank by the number of number operators $\hat{n}$ removed. The total number of basic diagrams to be evaluated is increased by unitary decomposition but the most complicated ones are reduced in number.

Let us return once again to product operators. We shall first assume that the unitary decomposition is carried out only on the individual elementary operators. This is far easier since the maximum particle rank is generally much lower and the possible types are quite limited compared with product operators. Let us assume that each elementary operator is already decomposed, first according to particle rank and then to unitary rank, in the following manner

$$\hat{O}_i = \sum_{k_i} \hat{O}_i(k_i, \mu_i) = \sum_{k_i} \sum_{\nu_i} \binom{\hat{n} - \nu_i - \mu_i}{k_i - \nu_i} \hat{O}_{k_i}^{\nu_i}(\nu_i, \mu_i). \tag{34}$$

Since $\hat{O}_{k_i}^{\nu_i}(\nu_i, \mu_i)$ is the minimum particle-rank member of the chain given in eq. (27), we can drop the redundant superscript ν_i here to simplify the notation. The product operator then takes the form

$$\hat{O} = \prod_{i=1}^q \hat{O}_i = \sum_{\text{all } \nu k} \prod_{i=1}^q \binom{\hat{n} - \nu_i - \mu_i}{k_i - \nu_i} \hat{O}_{k_i}(\nu_i, \mu_i). \tag{35}$$

Note that the number operator $\hat{n}$ in the product appears sandwiched between different $\hat{O}_{k_i}(\nu_i, \mu_i)$ operators.

Since each elementary operator may not be number-conserving by itself, the number operator $\hat{n}$ cannot be replaced simply by m, the number of active particles in the space, except when it occurs at either end of the product chain,

$$\binom{\hat{n} - \nu_1 - \mu_1}{k_1 - \nu_1} \hat{O}_{k_1}(\nu_1, \mu_1) \binom{\hat{n} - \nu_2 - \mu_2}{k_2 - \nu_2} \hat{O}_{k_2}(\nu_2, \mu_2)$$

$$\times \cdots \times \binom{\hat{n} - \nu_q - \mu_q}{k_q - \nu_q} \hat{O}_{k_q}(\nu_q, \mu_q).$$

This is especially true for configuration averaging, to be discussed in the next chapter, where number operators and number conservations must be introduced for each individual orbit.

Recalling that $\mu = (N_A - N_B)/2$, we have the *commutation* relation

$$\hat{O}(p, \mu)\, \hat{n} = (\hat{n} - 2\mu)\, \hat{O}(p, \mu),$$

for arbitrary particle rank p. This is due to the fact that, for $\mu \neq 0$, an operator $\hat{O}(p, \mu)$ creates 2μ particles more than it annihilates. In general,

$$\hat{O}(p, \mu') \binom{\hat{n} - \nu - \mu}{k - \nu} = \binom{\hat{n} - \nu - \mu - 2\mu'}{k - \nu} \hat{O}(p, \mu'). \tag{36}$$

Using this relation, we can now systematically move all the binomial coefficients involving the number operator to the front of the product chain, and eq. (35) then becomes

$$\hat{O} = \sum_{\substack{\text{all} \\ \nu k \mu}} \binom{\hat{n} - \nu_1 - \mu_1}{k_1 - \nu_1} \binom{\hat{n} - \nu_2 - \mu_2 - 2\mu_1}{k_2 - \nu_2} \binom{\hat{n} - \nu_3 - \mu_3 - 2(\mu_1 + \mu_2)}{k_3 - \nu_3}$$

$$\cdots \binom{\hat{n} - \nu_q - \mu_q - 2(\mu_1 + \mu_2 + \cdots + \mu_{q-1})}{k_2 - \nu_2} \prod_{i=1}^{q} \hat{O}_{k_i}(\nu_i, \mu_i), \tag{37}$$

a form which can be used for trace evaluations.

In the m-particle space, the number operators can now be taken out of the trace by replacing them by m,

$$\ll \hat{O} \gg^{(m)} = \sum_{\text{all } \nu k} \left\{ \prod_{i=1}^{q} \binom{m - \nu_i - \mu_i - 2\sum_{j=1}^{i-1} \mu_j}{k_i - \nu_i} \right\}$$

$$\times \langle \hat{O}_{k_1}(\nu_1, \mu_1) \cdots O_{k_q}(\nu_q, \mu_q) \rangle^{(m)}. \tag{38}$$

For the trace to be non-vanishing, it is necessary that

$$\mu \equiv \sum_{i=1}^{q} \mu_i = 0.$$

The total number of contractions required to fully contract the product of q unitarily decomposed operators $\hat{O}_{k_1}(\nu_1,\mu_1)\cdots\hat{O}_{k_q}(\nu_q,\mu_q)$ in eq. (38) is $\nu \equiv \sum_{i=1}^{q} \nu_i$. It is smaller than the maximum particle rank by the number of self-contractions (or number operators) in the diagram. Eq. (20) can now be used to evaluate the trace remaining on the right hand side of eq. (38),

$$\ll \hat{O}_{k_1}(\nu_1,\mu_1)\cdots\hat{O}_{k_q}(\nu_q,\mu_q) \gg^{(m)}$$

$$= \delta_{\mu 0} \sum_{t=0}^{\nu} \binom{N-\nu}{m-t} O_t^{\nu}(k_1\nu_1\mu_1,\ldots,k_q\nu_q\mu_q), \qquad (39)$$

where the upper limit of the summation is the lesser of m and ν, and $O_t^{\nu}(k_1\nu_1\mu_1,\ldots,k_q\nu_q\mu_q)$ is the fully contracted quantity defined in eq. (18).

Unitary decomposition is not restricted to elementary operators. Since only the unitary scalar part of an operator is invariant under a rotation of the single-particle basis, it is the only part that can contribute to the trace which obviously must also be invariant under such a rotation. Hence if a part of an operator has only a single unitary rank (k,μ), only the $(k,-\mu)$ part of the remainder of the operator can be coupled to it to form a scalar, and is therefore the only part that can contribute to the trace.

It is possible to take advantage of this point in calculating the trace of a product operator. We can divide a product of q elementary operators into two groups $\hat{\mathcal{F}}$ and $\hat{\mathcal{G}}$ such that

$$\hat{\mathcal{F}} \equiv \prod_{i=1}^{r} \hat{O}_i, \qquad \text{and} \qquad \hat{\mathcal{G}} \equiv \prod_{i=r+1}^{q} \hat{O}_i.$$

Each part can be decomposed according to particle and unitary rank. Since only the unitary scalar product of $\hat{\mathcal{F}}$ and $\hat{\mathcal{G}}$ can contribute to the trace, we have

$$\ll \hat{O} \gg^{(m)} = \sum_{k,k'} \sum_{\nu} \ll \hat{\mathcal{F}}_k^{\nu}(k,\mu)\, \hat{\mathcal{G}}_{k'}^{\nu}(k',-\mu) \gg^{(m)}. \qquad (40)$$

However, not many realistic applications have been carried out so far using this method. In order to take advantage of eq. (40) in numerical calculations, computer codes must be developed so that particle and unitary rank decompositions of a product of elementary operators can be carried out easily.

V.7   Particle-hole symmetry

In a space made up of a finite number of single-particle states, a many-particle state can be referred to either by the number of particles m in the space or by the number of holes $(N-m)$, i.e., the additional number of particles required to fill up all the single-particle states. More precisely, for each particle state $|m\Gamma Mx\rangle$, where x represents all the additional labels except Γ, the angular momentum, and M, the projection of Γ on the quantization axis, we can define a complimentary hole state $|(N-m), \Gamma, -M, x_c\rangle$ by

$$\left(|m\Gamma Mx\rangle \times |(N-m), \Gamma, -M, x_c\rangle\right)_{\text{antisymmetrized}} \equiv |N\rangle, \qquad (41)$$

where $|N\rangle$ is the closed shell state with all the single-particle states occupied. The N-particle space has only one state since there is only one way to fill all the single-particle states.

To evaluate a matrix element, we have therefore the choice either to carry out the calculation in terms of particles or in terms of holes. Let us define the particle-hole transformation for an operator by

$$\langle m\Gamma x \| \hat{O} \| m'\Gamma'x' \rangle \equiv \langle (N-m')\Gamma'x'_c \| \tilde{O} \| (N-m)\Gamma x_c \rangle. \qquad (42)$$

For single-particle creation and annihilation operators, the transformations are simply

$$\tilde{A}^\rho_\mu = B^\rho_{-\mu}, \qquad \tilde{B}^\rho_\mu = A^\rho_{-\mu}. \qquad (43)$$

Except for possible phase factors, these are identical to the transformation between an operator and its hermitian adjoint given by eqs. (C-11, 12). This similarity enables us to use particle-hole symmetry to simplify trace calculations.

We shall carry out the discussion of particle-hole transformations for the more complicated operators in terms of normal-ordered products,

$$\hat{O} = A^r A^s \cdots A^w B^{r'} B^{s'} \cdots B^{v'}, \qquad (44)$$

and omit the projection quantum number to simplify the notation. The particle-hole transformation of this operator follows from eq. (43)

$$\tilde{O} = B^r B^s \cdots B^w A^{r'} A^{s'} \cdots A^{v'}. \qquad (45)$$

On the other hand, the hermitian adjoint of $\hat{O}$ is given by

$$\overline{O} = (-1)^{2(r+s+\cdots+w)} A^{v'} \cdots A^{s'} A^{r'} B^w \cdots B^s B^r. \qquad (46)$$

To go from eq. (45) to (46) we must reorder the single-particle operators using commutation relations. In general this is complicated by the commutators produced in the process. However, for a fully contracted operator, defined by eq. (28), all the commutators must vanish and we are left only with a phase factor given by the total number of interchanges between adjacent single-particle operators. This number can be evaluated quite simply. An arbitrary operator of unitary rank (ν, μ) can always be written in a form with a total of 2ν single particle operators, $(\nu + \mu)$ creation operators, and $(\nu - \mu)$ annihilation operators. In order to reverse the order of 2ν quantities, we must move the leftmost one to the right across $(2\nu - 1)$ quantities, the second one across $(2\nu - 2)$, and so on, until the last one which does not have to be moved. The total number of interchanges is therefore

$$(2\nu - 1) + (2\nu - 2) + \cdots + 0 = \nu(2\nu - 1).$$

Hence we have

$$\tilde{O} = (-1)^h \overline{O}, \tag{47}$$

where

$$h = \nu(2\nu - 1) + 2(r + s + \cdots + w).$$

Note that, in the phase factor, the Latin indices, $r, s, \ldots, w$, stand for angular momenta and isospin of the orbits involved while the Greek index 2ν is simply the total number of single-particle creation and annihilation operators in the product operator.

For a product operator formed of q elementary operators,

$$\hat{O} = \hat{O}_1 \hat{O}_2 \cdots \hat{O}_q,$$

a particle-hole transformation leads to

$$\tilde{O} = \tilde{O}_1 \tilde{O}_2 \cdots \tilde{O}_q.$$

On the other hand,

$$\overline{O} = \overline{O}_q \cdots \overline{O}_2 \overline{O}_1.$$

Again, in general, $\tilde{O}$ and $\overline{O}$ are not simply related because of the commutators produced when we rearrange the order of the elementary operators. However, for traces, it often happens that

$$\ll \hat{O}_1 \hat{O}_2 \cdots \hat{O}_q \gg^{(m)} = \ll \hat{O}_q \cdots \hat{O}_2 \hat{O}_1 \gg^{(m)} . \tag{48}$$

For such traces, we have

$$\ll \hat{O}_1 \hat{O}_2 \cdots \hat{O}_q \gg^{(m)} = \ll \tilde{O}_q \cdots \tilde{O}_2 \tilde{O}_1 \gg^{(N-m)}$$

$$= (-1)^h \ll \overline{O}_q \cdots \overline{O}_2 \overline{O}_1 \gg^{(N-m)}$$

$$= (-1)^h \ll \hat{O}_1 \hat{O}_2 \cdots \hat{O}_q \gg^{(N-m)} , \tag{49}$$

where

$$h = \sum_{i=1}^{q} \nu_i(2\nu_i - 1).$$

It is assumed here that each elementary operator $\hat{O}_i$ is unitarily decomposed and has ranks (ν_i, μ_i).

Using eq. (49), we can rewrite eq. (39) for product operators satisfying eq. (48) in the form

$$\ll \hat{O} \gg^{(m)} = \sum_{t=0}^{[\frac{\nu}{2}]} \left[\binom{N-\nu}{m-t} + (-1)^h \binom{N-\nu}{m-\nu+t} \right]$$

$$\times \frac{1}{1+\delta_{t,\frac{\nu}{2}}} O_t^\nu (k_1\nu_1\mu_1, \ldots, k_q\nu_q\mu_q), \quad (50)$$

where the upper limit of the summation is

$$\left[\frac{\nu}{2} \right] = \begin{cases} \frac{\nu}{2} & \text{for } \nu \text{ even,} \\ \frac{\nu-1}{2} & \text{for } \nu \text{ odd,} \end{cases}$$

unless $m < \left[\frac{\nu}{2} \right]$, when the upper limit is simply m. The major advantage of eq. (50) is that the summation over t, and hence the number of different diagrams, is reduced to half of that in eq. (39), with a corresponding saving in the amount of computation.

V.8   Traces of powers of the Hamiltonian

We shall illustrate some of the important features of the diagrammatic method by working out examples involving $\langle H^r \rangle^{(m)}$ for power r up to four. These averages are the low-order moments of a scalar density distribution in the m-particle space. For simplicity we shall not use the particle-hole symmetry here. The Hamiltonian, given by eq. (IV-12), is a number-conserving operator, and therefore has $\mu = 0$. The maximum particle rank is two, and the operator can be decomposed according to unitary rank into $\nu = 0, 1,$ and 2 parts,

$$H = H(0,0) + H(1,0) + H(2,0).$$

The unitary rank-zero part is given by

$$H(0,0) = \bar{\epsilon}\hat{n} + \overline{W}\frac{\hat{n}(\hat{n}-1)}{2}, \quad (51)$$

where $\bar{\epsilon}$ is defined by eq. (IV-7) and $\overline{W}$ by eq. (IV-9). The $\nu = 1$ part is made up of the traceless single-particle energies $\tilde{\epsilon}_r$ of eq. (23), the off-diagonal single particle energy ϵ_{rs} for $r \neq s$, and the unitary rank one part of the two-body Hamiltonian $H(2)$ given by eq. (33). We can write $H(1,0)$ in the form

$$H(1,0) = \sum_{rs} \mathcal{E}_{rs}[r]^{\frac{1}{2}} \left(A^r \times B^s \right)^0, \quad (52)$$

where

$$\mathcal{E}_{rs} = \tilde{\epsilon}_r \delta_{rs} + \epsilon_{rs} + (\hat{n} - 1)\overline{W}_{rs}. \tag{53}$$

Note that, by definition, the off-diagonal single-particle energy $\epsilon_{rs} = 0$ for $r = s$, and the factor $(\hat{n} - 1)$ in the contribution from the two-body part is to ensure that the matrix element vanishes in the space of one particle. If $H(1,0)$ is to be used only in conjunction with other number-conserving elementary operators, as we shall be doing in this section, the number operator $\hat{n}$ in the definition of $\mathcal{E}_{rs}$ may be replaced by m, the particle number. This is however not true in general since $H(1,0)$ may be operating in intermediate states with different numbers of particles than in the space in which the trace is taken. The unitary rank-two part comes only from the two-body part and is given by

$$H(2,0) = -\sum_{\substack{rstu \\ JT}} \varsigma_{rs}\varsigma_{tu}[JT]^{\frac{1}{2}} \mathcal{W}_{rstu}^{JT}\left((A^r \times A^s)^{JT} \times (B^t \times B^u)^{JT}\right)^{00}, \tag{54}$$

where $\mathcal{W}_{rstu}^{JT}$, the two-body matrix element for the $\nu = 2$ part of H given by eq. (D-31), comes from the two-body matrix element W_{rstu}^{JT} after subtracting out the unitary rank zero and one parts.

Only the unitary rank-zero part of H can contribute to the centroid, and we obtain the result

$$\overline{E}_m = \langle H \rangle^{(m)} = m\bar{\epsilon} + \frac{m(m-1)}{2}\overline{W}, \tag{55}$$

with the aid of eq. (51).

For the trace of H^2, we need only the unitary rank-zero part of the product of two H's. Hence,

$$\langle H^2 \rangle^{(m)} = \langle H^2(0,0) + H^2(1,0) + H^2(2,0) \rangle^{(m)}.$$

The first term on the right-hand side, shown in Figs. V-3a and b, is the square of the unitary rank-zero part of H already calculated in eq. (55). The discussion is therefore simpler if we eliminate this term by dealing instead with the variance given by

$$\sigma_m^2 = \langle (H - \langle H \rangle^{(m)})^2 \rangle^{(m)} = \langle H^2(1,0) + H^2(2,0) \rangle^{(m)}. \tag{56}$$

This is equivalent to discarding the contributions from the disconnected diagram in H^2.

The contribution from $H(1,0)$ can be evaluated in the following way. A unitary rank-one operator is composed of one creation operator and one annihilation operator. Except for self-contractions already excluded by eliminating the unitary rank-zero part, there is only one way to contract the product of two $\nu = 1$ operators, the creation operator of the first one with the annihilation operator of the second one, and *vice versa*. They are represented by the diagrams shown in Fig. V-3c for the product of the $k = 1$ parts of the two Hamiltonians, in Fig. V-3d for the product of the $k = 1$ part of one Hamiltonian and the $k = 2$ part of the other, and in Fig. 3e for the product of the $k = 2$ parts from both. Since $\mathcal{E}_{rs}$ defined in eq. (53) consists of the contributions of the $\nu = 1$ parts from both the $k = 1$ and 2 parts of the Hamiltonian, the contributions of all three diagrams to the fully contracted quantity O_t^p, defined in eq. (18), is given by

$$O_1^2 = \sum_{rs} [r] \, \mathcal{E}_{rs} \, \mathcal{E}_{sr} \, \delta_{rs}^{\text{a.m.}}. \tag{57}$$

The propagator for this term, $\binom{N-2}{m-1}$, is obtained from the fact that there are two contractions $(p = 2)$ and one of them is a left-contraction $(t = 1)$. From eq. (20), we obtain

$$\ll H^2(1,0) \gg^{(m)} = \binom{N-2}{m-1} \sum_{rs} [r] \, \mathcal{E}_{rs} \, \mathcal{E}_{sr} \, \delta_{rs}^{\text{a.m.}}, \tag{58}$$

for the contribution to the trace from $H^2(1,0)$.

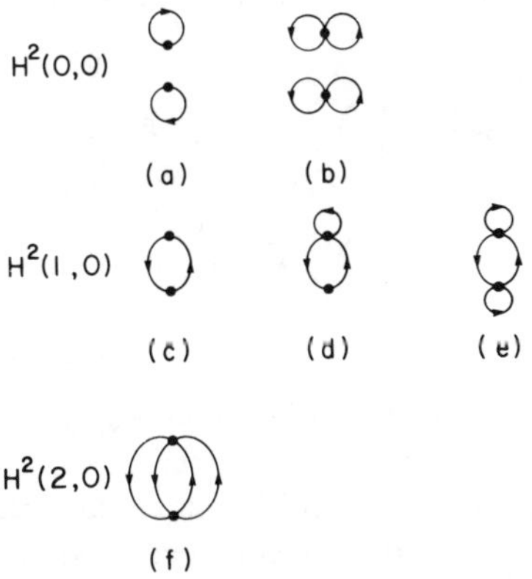

$H^2(0,0)$ (a) (b)

Fig. V-3. Diagrams associated with the contraction of H^2 with each Hamiltonian operator decomposed according to unitary rank.

$H^2(1,0)$ (c) (d) (e)

$H^2(2,0)$ (f)

There are three different ways to contract the product of two two-body operators and these are shown in Figs. V-3 b, e, and f. The first one (Fig. V-3b) is a disconnected diagram involving only self-contractions. This is then the $\nu = 0$ part of $H^2(2)$ and is excluded from the variance. Fig. V-3e involves one self-contraction in each of the two operators, and is therefore the $\nu = 1$ contribution of $H(2)$ already included in eq. (58). The third diagram, Fig. V-3f, consists of the purely $\nu = 2$ part of the Hamiltonian. We shall see that, because of the symmetry in two-body matrix element given by eq. (21), there is only one way to perform the contractions.

There are two single-particle creation operators and two single-particle annihilation operators associated with each of the two $\nu = 2$ operators. Since self-contraction is excluded, we must contract the two single-particle creation operators, say r_1 and s_1, of the first $H(2,0)$ with the two annihilation operators, t_2 and u_2, of the second $H(2,0)$. Similarly, the two annihilation operators, t_1 and u_1, of the first $H(2,0)$ must be contracted with the two creation operators, r_2 and s_2, of the second one. There are four different ways to carry out the four contractions, which may be represented as

[1] $\quad (r_1 t_2),\ (s_1 u_2),\ (t_1 r_2),\ (u_1 s_2),$

[2] $\quad (r_1 t_2),\ (s_1 u_2),\ (t_1 s_2),\ (u_1 r_2),$

[3] $\quad (r_1 u_2),\ (s_1 t_2),\ (t_1 r_2),\ (u_1 s_2),$

[4] $\quad (r_1 u_2),\ (s_1 t_2),\ (t_1 s_2),\ (u_1 r_2).$

These four ways are, however, related to each other. For example, the contraction [2] for the term involving $W^{\Gamma}_{r_1 s_1 t_1 u_1}(1)$ and $W^{\Gamma}_{t_2 u_2 r_2 s_2}(2)$ is same as [1] involving $W^{\Gamma}_{r_1 s_1 t_1 u_1}(1)$ and $W^{\Gamma}_{u_2 t_2 r_2 s_2}(2)$. Because of the symmetry between two-body matrix elements expressed by eq. (21), they give identical contributions to the trace. Similarly, [3] and [4] are also the same as [1] by the symmetry between W^{Γ}_{rstu} and W^{Γ}_{srtu}. A single diagram is therefore adequate to represent all the contributions of the $\nu = 2$ part of the Hamiltonian to the variance. The value of the fully contracted quantity is

$$O_2^4 = \frac{1}{4} \sum_{\substack{rstu \\ \Gamma}} [\Gamma] W^{\Gamma}_{rstu} W^{\Gamma}_{turs} (1 + \delta_{rs})(1 + \delta_{tu})$$

$$= \frac{1}{4} \sum_{\substack{rstu \\ \Gamma}} [\Gamma] \{W^{\Gamma}_{rstu}\}^2 (1 + \delta_{rs})(1 + \delta_{tu}) = \sum_{\substack{r \leq s \\ t \leq u \\ \Gamma}} [\Gamma] \{W^{\Gamma}_{rstu}\}^2, \quad (59)$$

where we have made use of $W^{\Gamma}_{rstu} = W^{\Gamma}_{turs}$ and the symmetries of the two-body matrix elements given in eq. (21).

Since $p = 4$ and $t = 2$ for this term, the propagator is $\binom{N-4}{m-2}$. Together with the results given by eqs. (58) and (59) we have the complete result for the variance,

$$\sigma_m^2 = \binom{N}{m}^{-1} \left\{ \binom{N-2}{m-1} \sum_{rs} [r] \mathcal{E}_{rs} \mathcal{E}_{sr} \delta_{rs}^{\text{a.m.}} + \binom{N-4}{m-2} \sum_{\substack{r \leq s \\ t \leq u \\ \Gamma}} [\Gamma] \{W^{\Gamma}_{rstu}\}^2 \right\}.$$

$$(60)$$

As illustrations, the centroids and widths for $m = 2$ to 23 are given in Table V-1 for two commonly used ds-shell interactions, the Chung-Wildenthal (Wildenthal and Chung, 1979) interaction obtained by fitting ds-shell experimental data and the Kuo (1967) interaction obtained by renormalizing a realistic free nucleon-nucleon interaction.

Table V-1. Centroids and widths in (ds)-space
calculated with Chung-Wildenthal and Kuo interactions

	CW$_{part.}$		Kuo			CW$_{hole}$		Kuo	
m	C	σ	C	σ	m	C	σ	C	σ
2	−5.46	3.99	−5.74	3.95	13	206.60	10.93	−117.20	12.05
3	−9.41	5.54	−10.29	5.43	14	218.70	10.76	−134.00	11.91
4	−14.17	6.94	−15.95	6.75	15	230.30	10.45	−152.00	11.62
5	−19.75	8.19	−22.73	7.92	16	241.40	10.01	−171.10	11.18
6	−26.15	9.30	−30.63	8.95	17	251.90	9.45	−191.30	10.60
7	−33.36	10.26	−39.65	9.83	18	261.90	8.75	−212.60	9.86
8	−41.38	11.07	−49.78	10.56	19	271.40	7.92	−235.00	8.96
9	−50.22	11.73	−61.03	11.15	20	280.30	6.94	−258.50	7.90
10	−59.87	12.25	−73.40	11.60	21	288.70	5.82	−283.20	6.66
11	−70.33	12.62	−86.88	11.90	22	296.50	4.52	−309.00	5.22
12	−81.61	12.85	−101.50	12.05	23	303.90	2.97	−335.90	3.47

For the higher-order moments, we shall ignore the $\nu = 0$ part of the Hamiltonian. This is equivalent to calculating the central moments as we have already done for the average of H^2 in terms of the variance. For the skewness of the distribution we need $\langle H^3 \rangle^{(m)}$. In terms of the unitarily decomposed Hamiltonian,

$$
\begin{aligned}
H^3 = \ & H^3(1,0) \\
& + H^2(1,0)H(2,0) + H(1,0)H(2,0)H(1,0) + H(2,0)H^2(1,0) \\
& + H(1,0)H^2(2,0) + H(2,0)H(1,0)H(2,0) + H^2(2,0)H(1,0) \\
& + H^3(2,0).
\end{aligned}
\tag{61}
$$

In scalar averaging we have $\mu = 0$ for Hamiltonians. As a result, we can apply the cyclic permutation invariance of the trace of a product of operators and obtain

$$
\begin{aligned}
\langle H^2(1,0)H(2,0) \rangle^{(m)} &= \langle H(1,0)H(2,0)H(1,0) \rangle^{(m)} \\
&= \langle H(2,0)H^2(1,0) \rangle^{(m)},
\end{aligned}
\tag{62a}
$$

and

$$
\begin{aligned}
\langle H(1,0)H^2(2,0) \rangle^{(m)} &= \langle H(2,0)H(1,0)H(2,0) \rangle^{(m)} \\
&= \langle H^2(2,0)H(1,0) \rangle^{(m)}.
\end{aligned}
\tag{62b}
$$

With this result, we obtain the expression

$$
\langle H^3 \rangle^{(m)} = \langle \left(H^3(1,0) + 3H^2(1,0)H(2,0) + 3H(1,0)H^2(2,0) + H^3(2,0) \right) \rangle^{(m)},
\tag{63}
$$

Fig. V-4. Diagrams associated with the contraction of three Hamiltonian operators decomposed according to unitary ranks. (a) and (b) are for $\nu = 1$ parts, (c) is for the product of two $\nu = 1$ and one $\nu = 2$ operator, (d) and (e) are for one $\nu = 1$ and two $\nu = 2$ operators, whereas (f) and (g) are for $\nu = 2$ operators only.

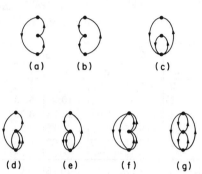

for the average scalar trace of H^3 given in eq. (61).

The first term is the simplest one to evaluate since it involves only the unitary rank-one part of the Hamiltonian. There are two different ways to contract the three $\nu = 1$ operators and they are diagrammatically represented by Figs. V-4a and b. As indicated by the arrows in the figures, these two diagrams differ only in the directions of the contractions, *i.e.*, whether the single-particle creation operator is to the right or left of the annihilation operator it contracts with. The two diagrams are therefore related and have the same O_t^p value. However, the propagators are different since Fig. V-4a has $(p, t) = (3, 2)$, whereas Fig. V-4b has $(p, t) = (3, 1)$. For convenience of comparison with the configuration case to be discussed in the next chapter, it is better to write down the more general result, allowing the possibility that the three Hamiltonians may be different,

$$\ll H(1,0)_1 H(1,0)_2 H(1,0)_3 \gg^{(m)}$$

$$= \left\{ \binom{N-3}{m-1} + \binom{N-3}{m-2} \right\} \sum_{rst} [r] \mathcal{E}_{rs}(1) \, \mathcal{E}_{st}(2) \, \mathcal{E}_{tr}(3) \, \delta_{rs}^{\text{a.m.}} \cdot \delta_{st}^{\text{a.m.}}, \quad (64)$$

where $\mathcal{E}_{rs}(i)$ is given by eq. (53) and $i = 1, 2$, and 3, indicates which of the three Hamiltonians it comes from.

Only one diagram is associated with the contraction of the product of two $\nu = 1$ and one $\nu = 2$ operator and this is shown in Fig. V-4c. There are four contractions ($p = 4$), two left-contractions ($t = 2$) and two right-contractions. The diagram is not changed by applying any of the symmetry relations in reverse. Consequently we have only one term for the scalar trace of $H^2(1,0)H(2,0)$, given by

$$\ll H(1,0)_1 H(1,0)_2 H(2,0) \gg^{(m)}$$

$$= -\binom{N-4}{m-2} \sum_{rstu} \mathcal{E}_{rt}(1) \mathcal{E}_{su}(2) \beta_{rtsu}^0 \delta_{rt}^{\text{a.m.}} \cdot \delta_{su}^{\text{a.m.}}, \quad (65)$$

where the multipole coefficient for the unitary rank $\nu = 2$ part of the Hamiltonian is defined by

$$\beta^{\Delta}_{rtsu} = \sum_{\Gamma} (-1)^{s+t+\Gamma+\Delta} [\Delta]^{\frac{1}{2}} [\Gamma] \left\{ \begin{matrix} r & s & \Gamma \\ u & t & \Delta \end{matrix} \right\} W^{\Gamma}_{rstu}, \qquad (66)$$

in analogy with eq. (D-24).

The two contractions of $H(1,0)H^2(2,0)$, one with $(p,t) = (5,2)$ and the other with $(p,t) = (5,3)$, are shown in Figs. V-4d and e. They are topologically identical and can be transformed into each other by changing the directions of the contractions (or by inverting the positions of the three Hamiltonians). Consequently,

$$\ll H(1,0)H^2(2,0) \gg^{(m)} = \left\{ \binom{N-5}{m-2} + \binom{N-5}{m-3} \right\}$$

$$\times \frac{1}{2} \sum_{\substack{rstuw \\ \Gamma}} [\Gamma] \mathcal{E}_{rs} W^{\Gamma}_{stuw}(1) W^{\Gamma}_{uwrt}(2) \delta^{\text{a.m.}}_{rs}, \qquad (67)$$

since the contributions of these two diagrams differ only in their propagators.

For the $H^3(2,0)$ term, there are two topologically distinct diagrams, and they are given in Figs. V-4f and g. The former represents the two possible ways to contract the three $\nu = 2$ operators corresponding to $(p,t) = (6,2)$ and $(6,4)$. Again, these two contractions differ only in the directions of the contractions. Fig. V-4g is, however, different and has $(p,t) = (6,3)$. The complete result for the contribution of $H^3(2,0)$ is then

$$\ll H(2,0)_1 H(2,0)_2 H(2,0)_3 \gg^{(m)}$$

$$= \left\{ \binom{N-6}{m-2} + \binom{N-6}{m-4} \right\} \frac{1}{8} \sum_{\substack{rstu \\ xy;\Gamma}} [\Gamma] W^{\Gamma}_{rstu}(1) W^{\Gamma}_{tuxy}(2) W^{\Gamma}_{xyrs}(3)$$

$$- \binom{N-6}{m-3} \sum_{\substack{rstu \\ xy;\Delta}} \frac{(-1)^{r-t-\Delta}}{[\Delta]^{\frac{1}{2}}} \beta^{\Delta}_{rtsu}(1) \beta^{\Delta}_{xysu}(2) \beta^{\Delta}_{xytr}(3). \qquad (68)$$

This result was also derived by Ayik and Ginocchio (1974); here, the derivation is carried out on a computer using algebraic codes (Chang and Wong, 1979, 1980) based on the method described in part in Appendix E.

For simplicity, we shall give only the result of the $\nu = 2$ part for H^4. There are altogether 42 different ways to contract the 16 single-particle creation and annihilation operators. All 42 diagrams can be generated from the six basic diagrams shown in Fig. V-5. Let us first write down the

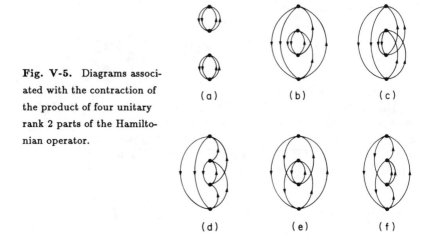

Fig. V-5. Diagrams associated with the contraction of the product of four unitary rank 2 parts of the Hamiltonian operator.

(a)   (b)   (c)

(d)   (e)   (f)

form of O_t^p for each of the basic diagrams. The first one, Fig. V-5a, is a disconnected diagram and the fully contracted quantity is given by

$$O_4^8(1234)_1 = \frac{1}{16} \sum_{\substack{rstu;\Gamma_1 \\ wxyz;\Gamma_2}} [\Gamma_1\Gamma_2] W_{rstu}^{\Gamma_1}(1) W_{wxyz}^{\Gamma_2}(2) W_{yzwx}^{\Gamma_2}(3) W_{turs}^{\Gamma_1}(4). \quad (69a)$$

It is essentially the product of two traces, each involving two $H(2,0)$'s. From the positions of the indices r, s, t, u, w, x, y, and z, we can see that eq. (69a) represents the contraction of $H(2,0)_1$ with $H(2,0)_4$, and $H(2,0)_2$ with $H(2,0)_3$, where the subscripts 1, 2, 3, and 4 indicate the order of the four $H(2,0)$'s as they appear on the right-hand side of eq. (69a). Two other possibilities must also be considered. Instead of contracting operators 1 with 4 and 2 with 3, we can contract operators 1 with 3 and 2 with 4. It is easy to see that in this case the form of the expression for O_t^p is $O_4^8(1243)_1$, the same as eq. (69a) except for a permutation of the operator indices 3 and 4. Similarly, we have the contraction of operators 1 with 2 and 3 with 4, and the fully contracted quantity is given by $O_4^8(1342)_1$, which can also be written as $O_4^8(2341)_1$. In all three cases, we have $(p,t) = (8,4)$, and therefore they all have the same propagator $\binom{N-8}{m-4}$.

The expressions for O_t^p's for the other five basic diagrams can be written in the form

$$O_4^8(1234)_2 = -\frac{1}{4} \sum_{\substack{rstu;\Gamma_1 \\ wxyz;\Gamma_2}} \left[\frac{\Gamma_1\Gamma_2}{u} \right]$$

$$\times W_{rstu}^{\Gamma_1}(1)\, W_{wxyz}^{\Gamma_2}(2)\, W_{uzwx}^{\Gamma_2}(3)\, W_{tyrs}^{\Gamma_1}(4)\, \delta_{uy}^{\text{a.m.}}, \quad (69b)$$

$$O_4^8(1234)_3 = \frac{1}{16} \sum_{\substack{rstu;\Gamma \\ wxyz}} \mathcal{W}_{rstu}^{\Gamma}(1) \, \mathcal{W}_{wxyz}^{\Gamma}(2) \, \mathcal{W}_{tuwx}^{\Gamma}(3) \, \mathcal{W}_{yzrs}^{\Gamma}(4), \qquad (69c)$$

$$O_3^8(1234)_4 = -\frac{1}{2} \sum_{\substack{rstu;\Gamma \\ wxyz;\Delta}} [\Gamma] \left\{ \begin{array}{ccc} t & u & \Gamma \\ x & z & \Delta \end{array} \right\} \mathcal{W}_{rstu}^{\Gamma}(1) \beta_{xuyw}^{\Delta}(2) \beta_{tzyw}^{\Delta}(3) \mathcal{W}_{xzrs}^{\Gamma}(4),$$

$$(69d)$$

$$O_4^8(1234)_5 = \sum_{\substack{rstu;\Delta \\ wxyz}} \frac{1}{[\Delta]} \beta_{rtsu}^{\Delta}(1) \, \beta_{wyxz}^{\Delta}(2) \, \beta_{suxz}^{\Delta}(3) \, \beta_{rtwy}^{\Delta}(4), \qquad (69e)$$

$$O_4^8(1234)_6 = - \sum_{\substack{rstu;\Delta_1 \\ wxyz;\Delta_2}} (-1)^{s+x+\Delta_1+\Delta_2} \left\{ \begin{array}{ccc} z & x & \Delta_1 \\ u & s & \Delta_2 \end{array} \right\}$$

$$\times \beta_{rtsu}^{\Delta_1}(1) \, \beta_{xuyw}^{\Delta_2}(2) \, \beta_{zsyw}^{\Delta_2}(3) \beta_{zxrt}^{\Delta_1}(4). \qquad (69f)$$

To find the other contractions associated with each basic diagram, we must apply all the symmetry relations in reverse. The situation is somewhat more involved than for the disconnected diagram. The principle is, however, the same except that the process of arriving at the final result is more tedious. Here, we shall simply quote the results generated with a simple computer program to carry out the work. The complete result for the trace of $H^4(2,0)$ is

$$\ll H(2,0)_1 H(2,0)_2 H(2,0)_3 H(2,0)_4 \gg^{(m)}$$

$$= \sum_{\substack{rstu \\ wxyz}} \left[\binom{N-8}{m-4} \left\{ O_4^8(1234)_1 + O_4^8(1243)_1 + O_4^8(2341)_1 \right\} \right.$$

$$+ \left\{ \binom{N-8}{m-3} + \binom{N-8}{m-5} \right\} \left\{ O_4^8(3142)_2 + O_4^8(2431)_2 \right\}$$

$$- \binom{N-8}{m-4} \left\{ O_4^8(1234)_2 + O_4^8(4321)_2 + O_4^8(3421)_2 + O_4^8(2134)_2 \right.$$

$$\left. + O_4^8(1423)_2 + O_4^8(3241)_2 + O_4^8(3214)_2 + O_4^8(1432)_2 \right\}$$

$$+ \left\{ \binom{N-8}{m-2} + \binom{N-8}{m-6} \right\} O_4^8(2431)_3$$

$$+ \binom{N-8}{m-4} \left\{ O_4^8(4312)_3 + O_4^8(4321)_3 + O_4^8(2341)_3 + O_4^8(2314)_3 \right\}$$

$$- \left\{ \binom{N-8}{m-3} + \binom{N-8}{m-5} \right\} \left\{ O_3^8(1234)_4 + O_3^8(3142)_4 \right.$$

$$\left. + O_3^8(2341)_4 + O_3^8(4123)_4 \right\}$$

$$+ \binom{N-8}{m-4} \{O_3^8(4312)_4 + O_3^8(3241)_4 + O_3^8(1423)_4 + O_3^8(2314)_4\}$$

$$+ \binom{N-8}{m-4} \{O_4^8(4321)_5 + O_4^8(4231)_5 + O_4^8(3241)_5\}$$

$$+ \left\{ \binom{N-8}{m-3} + \binom{N-8}{m-5} \right\} O_4^8(1243)_6$$

$$- \binom{N-8}{m-4} \{O_4^8(2413)_6 + O_4^8(1324)_6 + O_4^8(3214)_6 + O_4^8(1432)_6\} \Bigg].$$

$$(70)$$

It is not difficult to check that each of the 42 terms in eq. (70) can be generated from the six basic diagrams using the permutations indicated. Essentially the same results were also given by Ayik and Ginocchio (1974).

Chapter VI

Configuration averaging

So far we have been dealing mainly with *scalar* averaging in which the strength in the entire m-particle space is described by a single distribution. The calculations involved are relatively simple since only one set of moments is required and each moment is the trace over all the states in the space. However, as mentioned in Chapter II, it is not always advantageous to use a single distribution since one would need the high-order moments to give the necessary accuracy in the ground state region. For a calculation restricted to the low-order moments, the accuracy can be improved by *partitioning* the space.

In this chapter, our primary concern is the strength distribution in the subspace of a configuration m. To obtain the low-order moments that characterize the smooth variations of the distribution, we shall calculate the average traces of product operators over all the states within the subspace of a configuration. We shall rely on the shorthand notation introduced in chapter IV of using a bold-faced letter to represent a set of quantities, one for each orbit, and we shall refer to the m-particle space as the *scalar* space.

VI.1 Diagrammatic method for configuration trace

As described in the previous chapter, the first step in a diagrammatic method of trace calculations is to find all the basic diagrams. The only difference between scalar and configuration averaging is that the orbit of each pair of contracting single-particle operators must be identified. The reason, as we shall see later, is that the propagator here depends on the number and type of contractions in each orbit.

The term *diagram* is commonly employed to mean two related things. The first is the picture identifying the contraction patterns, such as those shown in Figs. V-2 to V-5. In order to adapt these pictures to configuration averaging, we should in principle add an orbital index for each contraction. However, in general this is not done since it only tends to complicate the picture without adding anything useful at this stage. The second is the expression corresponding to the way to contract all the single particle operators. In other words, it is essentially the algebraic expression for the fully contracted quantity O_t^p defined in eq. (V-18). Examples of such expressions are given by eqs. (V-59) and (V-69). We note that, in scalar averaging, the summation over orbital indices can be regarded as part of the calculation of O_t^p, whereas in configuration averaging this summation is carried out later. As a result, the value of O_t^p in configuration averaging is a function of the orbits involved in the contraction. In either sense of the word, the diagrams involved in scalar and configuration averaging are the same provided the association with orbits is properly understood.

By the same token, except for the need of orbit identification, all the arguments leading up to eq. (V-20) for scalar averaging apply equally well to configuration averaging. As a result, in analogy to eq. (V-20), the starting expression for the configuration trace of operator $\hat{O}$ is

$$\ll \hat{O} \gg^{(m)} = \sum_{t=0}^{p} \binom{N-p}{m-t} O_t^p, \tag{1}$$

where O_t^p is the fully contracted value of the diagram defined in a similar manner as in eq. (V-18) for the scalar case. The maximum particle ranks of $\hat{O}$ in each orbit are given by $p \equiv \{p_1, p_2, \ldots, p_A\}$, which are also the maximum numbers of contractions in the orbits. Among the p contractions, the number of left-contractions is given by $t \equiv \{t_1, t_2, \ldots, t_A\}$. The shorthand notation for the propagator is defined by

$$\sum_{t=0}^{p} \binom{N-p}{m-t} \equiv \sum_{t_1=0}^{p_1} \sum_{t_2=0}^{p_2} \cdots \sum_{t_A=0}^{p_A} \binom{N_1 - p_1}{m_1 - t_1}\binom{N_2 - p_2}{m_2 - t_2} \cdots \binom{N_A - p_A}{m_A - t_A}. \tag{2}$$

For convenience, we can think of the propagator in the configuration case as made up of a product of propagators, one from each orbit, with the value for orbit r given by $\binom{N_r - p_r}{m_r - t_r}$. For an orbit without contractions, $p_r = t_r = 0$, the contribution to the propagator is simply the number of different ways the m_r particles can be put into the N_r single-particle states, the same factor as the contribution of the orbit to the dimension of the configuration subspace. As a result, the form of the propagator is usually simpler for an average trace.

Most of the symmetry properties used in Chapter V to reduce the number of basic diagrams associated with the trace of a particular product operator are not affected by changing to configuration subspace. The permutation symmetry between equivalent elementary operators is independent of the orbit in which a particular contraction takes place, and is therefore unaffected by partitioning the m-particle space. Furthermore, since our elementary operators are defined in spaces organized into orbits, both the symmetry in the defining matrix elements and the particle-hole symmetry operate in the same way in configuration subspaces as in scalar spaces.

The unitary symmetry is however quite different. In configuration subspace, there is a number operator $\hat{n}_r$ for each of the active orbits; in contrast, there is only one number operator $\hat{n}$ in the scalar space. Under a unitary transformation among the N_r single-particle states in orbit r, the operator $\hat{n}_r$ (as well as $\hat{n}$) remains invariant. On the other hand, only $\hat{n}$ is invariant under a unitary transformation among all N single-particle states. As a result, we have unitary ranks (ν_r, μ_r) for each orbit in configuration averaging. The property of an operator under a unitary transformation of this more general form can also be used to simplify configuration trace calculations. In the shorthand notation, the unitary ranks of an operator in configuration space can be written as (ν, μ), which are related to the scalar ranks (ν, μ) by

$$ \nu = \sum_r \nu_r, \qquad \mu = \sum_r \mu_r, $$

as can be seen from the relation between (ν, μ) and the total number of creation and the total number of annihilation operators given by eqs. (V-24, 25).

Each elementary operator $\hat{O}_i$ can now be written as a sum over terms with definite unitary ranks in all the orbits. Analogous to eq. (V-34), we have the expansion

$$ \hat{O}_i = \sum_{k_i} O_i(k_i, \mu_i) = \sum_{k_i} \sum_{\nu_i} \binom{\hat{n} - \nu_i - \mu_i}{k_i - \nu_i} O_{k_i}^{\nu_i}(\nu_i, \mu_i), \qquad (3) $$

where the elementary operator $\hat{O}_i$ is first decomposed according to particle rank (k_i, μ_i), and each of the fixed-particle rank parts $O_i(k_i, \mu_i)$ is further decomposed according to the unitary rank in all the orbits. For a product of q elementary operators, we have the expression

$$ \hat{O} = \sum_{\text{all } \nu k} \prod_{i=1}^{q} \binom{\hat{n} - \nu_i - \mu_i}{k_i - \nu_i} O_{k_i}^{\nu_i}(\nu_i, \mu_i), \qquad (4) $$

which is analogous to eq. (V-35).

The product operator $\hat{O}$ is now expressed as a sum over terms each of which is a product made up of fixed particle- and unitary-rank parts from all the elementary operators. The summation quickly becomes rather cumbersome for anything other than simple product operators. Alternatively, we can decompose the entire product operator itself according to particle and unitary ranks rather than the individual elementary operators in it. This procedure is potentially even more attractive here than in scalar averaging. Since only the $(\nu, \mu) = (0, 0)$ part of $\hat{O}$ can contribute to the trace, we can discard all the other parts from the start. The only operation remaining in a unitary rank $(0,0)$ operator involves the (trivial) number operators and, as a result, unitary decomposition becomes the main part of the work to be carried out in a trace calculation. Although there is no fundamental difficulty in the method, this approach has seldom been used.

All the algebraic results of scalar averaging in Chapter V can essentially be carried over to configuration averaging by formally making the change from N to N, m to m, etc., since the expressions are similar for both types of averages if we make the necessary modifications in all the quantities that depend on orbits in the configuration case. The actual calculations are more involved since the propagator is now different for each term with a different set of orbital indices. Furthermore, since the unitary decomposition in the configuration case is more refined, more information is required to specify the decomposed operator.

The additional propagators and input quantities are, however, not the major problem in configuration space. Since the number of different diagrams remains unchanged for a given product operator when the m-particle space is subdivided according to configuration, the amount of computation to evaluate a particular trace increases by a factor roughly proportional to the number of orbits involved. For a particular configuration, it is seldom that all the active orbits actually enter into the calculation for a diagram, either because some orbits are not occupied or the unitary rank of the operator for the orbit is zero. When a large number of orbits is involved, the unitary rank and the occupancies of each orbit are usually low since their sums in the complete space are restricted. For such cases, the trace calculation is made simple by the low unitary ranks and particle numbers involved.

The main complication of configuration averaging comes simply from the large number of configurations in the space, but this is not a serious problem. Since each configuration distribution is specified by a separate set of moments, the total number of traces to be calculated becomes large. This can often be an advantage since by calculating the moments for each configuration we have increased the amount of information specifying the strength distribution. Furthermore, if our interest is restricted to a small region near the ground state, we need to evaluate only the moments of those distributions that provide the dominant contribution in the low excitation energy region.

VI.2   Configuration centroid and width

To obtain the density distribution of a configuration m, we need the average of H^r for small values of r in the subspace of m particles. The procedures are essentially the same as those in the scalar case described in Section V-8. The following discussion is given in part as illustration of the similarities and differences between the two averaging methods.

Again we start with the unitary decomposition of the Hamiltonian. In configuration subspaces the unitary rank-zero part, $H(0,0)$, has contributions from the induced single-particle energies $\overline{W}(r,s)$, defined by eq. (D-28), as well as from the (true) single-particle energies ϵ_r,

$$H(0,0) = \sum_r \epsilon_r \hat{n}_r + \sum_{r \leq s} \overline{W}(r,s) \frac{\hat{n}_r(\hat{n}_s - \delta_{rs})}{1 + \delta_{rs}}. \tag{5}$$

Since the induced single-particle energies come from the two-body part of the Hamiltonian, we have the Kronecker deltas in the expression to ensure that the $\overline{W}(r,s)$ term remains a two-body operator.

Only the unitary rank-zero part of H can contribute to the centroid, $\overline{E}_m$, of a configuration density distribution. Using eq. (5), we obtain

$$\overline{E}_m = \langle H \rangle^{(m)} = \sum_r m_r \epsilon_r + \sum_{r \leq s} \frac{m_r(m_s - \delta_{rs})}{1 + \delta_{rs}} \overline{W}(r,s). \tag{6}$$

The calculation is trivial since only number operators are involved.

The unitary rank-one part of the Hamiltonian can exist in the configuration case only if there are radially degenerate orbits, i.e., $j_r^{\pi_r} = j_s^{\pi_s}$ for $r \neq s$. This term also has contributions from both the one-body off-diagonal single-particle energy and two-body induced single-particle energy terms,

$$H(1,0) = \sum_{r \neq s} \epsilon_{rs} [r]^{\frac{1}{2}} (A^r \times B^s)^0 + \sum_{\substack{r \neq s \\ t}} \overline{W}(r,s;t) (\hat{n}_t - \delta_{rt}) [r]^{\frac{1}{2}} (A^r \times B^s)^0$$

$$\equiv \sum_{r \neq s} \tilde{\mathcal{E}}_{rs} [r]^{\frac{1}{2}} (A^r \times B^s)^0. \tag{7}$$

For later convenience, we have defined

$$\tilde{\mathcal{E}}_{rs} \equiv \epsilon_{rs} + \sum_t \overline{W}(r,s;t) (\hat{n}_t - \delta_{rt}). \tag{8}$$

Note that $\tilde{\mathcal{E}}_{rs}$ involves the operator $\hat{n}_t$; its value therefore depends on the number of particles in orbit t and is different from $\mathcal{E}_{rs}$ defined in eq. (V-53) for the scalar case. In eqs. (7) and (8), ϵ_{rs} is the off-diagonal single-particle energy, and the induced off-diagonal single-particle energy, $\overline{W}(r,s;t)$, is defined by eq. (D-30) in terms of two-body matrix elements. The δ_{rt} factor comes from the binomial coefficient in eq. (3) for the following reason. If $r \neq t$, we have $k_t = 1$ and $\nu_t = \mu_t = 0$, and the binomial coefficient for orbit t reduces to $\hat{n}_t$. If $r = t$, we have $k_t = 3/2$ and $\nu_t = \mu_t = 1/2$, and the binomial factor becomes $(\hat{n}_t - 1)$. This can also be seen in another way.

The original two-body operator is of the form $A^r A^t B^t B^s$. The $\nu = 1$ part, however, has the form $\hat{n}_t A^r B^s \rightarrow A^t B^t A^r B^s$. For $r = t$ ($\neq s$), the number operator now operates on a state of one more particle in orbit t, and this is corrected by the Kronecker delta.

After removing the $\nu = 0$ and $\nu = 1$ parts, the remaining two-body term takes on the form

$$H(2,0) = -\sum_{\substack{rstu \\ JT}} \varsigma_{rs}\varsigma_{tu}[JT]^{\frac{1}{2}} V_{rstu}^{JT}\big((A^r \times A^s)^{JT} \times (B^t \times B^u)^{JT}\big)^{00}, \quad (9)$$

where V_{rstu}^{JT} is given by eq. (D-32).

For the trace of H^2, we need the unitary rank-zero product of the two H's. As usual, we consider only the central moments and remove the $\nu = 0$ part from H, as we have done for the scalar case in Chapter V. The variance of the distribution is therefore a sum of the squares of the $\nu = 1$ and $\nu = 2$ parts of the Hamiltonian,

$$\sigma_m^2 = \langle (H - \langle H \rangle^{(m)})^2 \rangle^{(m)} = \langle H^2(1,0) + H^2(2,0) \rangle^{(m)}. \quad (10)$$

There is no contribution from the product of $H(1,0)$ and $H(2,0)$ since no unitary scalar can be formed.

Consider first the contribution of the square of the $\nu = 1$ part. The order of single-particle creation and annihilation operators is $A^r B^s A^{s'} B^{r'}$. Two contractions are needed to obtain the trace, a right-contraction between B^s and $A^{s'}$ requiring $s = s'$, and a left-contraction between A^r and $B^{r'}$ requiring $r = r'$. Since $r \neq s$ for $H(1,0)$, we have $(p_r, t_r) = (1,1)$, $(p_s, t_s) = (1,0)$, and $(p_w, t_w) = (0,0)$ for all $w \neq r \neq s$. From eq. (2), we obtain the propagator for this term as

$$\binom{N_1}{m_1}\binom{N_2}{m_2}\cdots\binom{N_r - 1}{m_r - 1}\cdots\binom{N_s - 1}{m_s}\cdots\binom{N_A}{m_A}.$$

This form will be simplified when we change to the average trace later.

The contribution from the induced single-particle energy is modified by the $(\hat{n}_t - \delta_{rt})$ factor in eq. (7). So long as $r \neq s \neq t$, the expectation value of $\hat{n}_t$ is simply m_t, the number of particles in orbit t for this configuration. This is also true in general for the $\hat{n}_t$ operator associated with the first (leftmost) $H(1,0)$ operator since it always operates in a space of m_t particles. However, when $\hat{n}_t$ comes from the second $H(1,0)$ operator, the value of $\hat{n}_t$ is $m_t + \delta_{st} - \delta_{rt}$ in general. This can easily be seen from the order of the single-particle operators for this term, $A^r B^s \hat{n}_t A^s B^r$. If $s = t$, $\hat{n}_t$ operates in the intermediate states of $m_t + 1$ particles, and if $r = t$ it operates in the intermediate states with $m_t - 1$ particles.

For the second $H(1,0)$ operator we are dealing with the contribution from $\overline{W}(s,r;t)$ instead of $\overline{W}(r,s;t)$ for the first $H(1,0)$ operator. Furthermore, for this $H(1,0)$, there is also the additional Kronecker delta term,

$-\delta_{st}$, from eq. (8). The complete factor associated with $\overline{W}(s,r;t)$ is there-fore $m_t + \delta_{st} - \delta_{rt} - \delta_{st} = m_t - \delta_{rt}$. The contribution to the variance from $H^2(1,0)$ is therefore given by

$$\ll H^2(1,0) \gg^{(m)} = \sum_{r \neq s} \binom{N_r - 1}{m_r - 1} \binom{N_s - 1}{m_s} \left\{ \prod_{\substack{w \neq r \\ w \neq s}} \binom{N_w}{m_w} \right\}$$

$$\times N_r \left\{ \epsilon_{sr} + (m_t - \delta_{rt})\overline{W}(r,s;t) \right\} \left\{ \epsilon_{sr} + (m_t - \delta_{rt})\overline{W}(s,r;t) \right\}. \quad (11)$$

The result can be simplified by dividing both sides of the equation by the dimension of the configuration, and we obtain the average trace of $H^2(1,0)$,

$$\langle H^2(1,0) \rangle^{(m)} = \sum_{r \neq s} \frac{m_r(N_s - m_s)}{N_r N_s} N_r \left\{ \epsilon_{sr} + (m_t - \delta_{rt})\overline{W}(r,s;t) \right\}^2, \quad (12)$$

where we have used the relations that $\epsilon_{rs} = \epsilon_{sr}$ and $\overline{W}(r,s;t) = \overline{W}(s,r;t)$ to obtain the final form.

The square of the $\nu = 2$ term is generally the more important one compared with the $\nu = 1$ term. This is due to the fact that the $\nu = 1$ term occurs between radially degenerate orbits only and these orbits are relatively few in number. As shown in Fig. V-3, there is only one independent way to contract the two $\nu = 2$ operators. Since each $H(2,0)$ is made up of two single-particle creation operators and two annihilation operators, four contractions are required, two left-contractions and two right-contractions. The fully contracted quantity O_t^p takes the form

$$O_{2(rstu)}^4 = \frac{1}{4} \sum_{\Gamma} [\Gamma] V_{rstu}^{\Gamma} V_{turs}^{\Gamma} (1 + \delta_{rs}) (1 + \delta_{tu})$$

$$= \frac{1}{4} \sum_{\Gamma} [\Gamma] \{V_{rstu}^{\Gamma}\}^2 (1 + \delta_{rs}) (1 + \delta_{tu}). \quad (13)$$

Except for the absence of summation over the indices $rstu$, the result is identical to that for the scalar case given by eq. (V-59).

We must now work out the propagator for the $H^2(2,0)$ term. Let us first assume that all four contractions occur in different orbits, i.e., $r \neq s \neq t \neq u$. In this case, we have $(p_r, t_r) = (p_s, t_s) = (1,1)$, $(p_t, t_t) = (p_u, t_u) = (1,0)$, and $(p_w, t_w) = (0,0)$ for all the other orbits. The propagator is then

$$\binom{N_r - 1}{m_r - 1} \binom{N_s - 1}{m_s - 1} \binom{N_t - 1}{m_t} \binom{N_u - 1}{m_u} \prod_{\substack{w \neq r, w \neq s \\ w \neq t, w \neq u}} \binom{N_w}{m_w}.$$

For the average trace, the dimension of the configuration space can be taken out, and only orbits involved in the contraction remain in the propagator.

Table VI-1. Dimensions, centroids, and widths of $(ds)^{12}$-configurations with Chung-Wildenthal particle interaction

$d_{5/2}$	$d_{3/2}$	$s_{1/2}$	d	C	σ	$d_{5/2}$	$d_{3/2}$	$s_{1/2}$	d	C	σ
12	0	0	1	-127.08	10.17	6	3	3	206976	-86.30	9.67
11	1	0	96	-117.70	9.37	6	2	4	25872	-94.91	8.73
11	0	1	48	-118.01	10.65	5	7	0	6336	-68.05	8.81
10	2	0	1848	-108.63	8.96	5	6	1	88704	-69.27	9.69
10	1	1	2112	-109.10	10.06	5	5	2	266112	-73.01	10.00
10	0	2	396	-112.08	10.69	5	4	3	221760	-79.27	9.67
9	3	0	12320	-99.88	8.80	5	3	4	44352	-88.04	8.73
9	2	1	24640	-100.50	9.79	4	8	0	495	-60.88	9.09
9	1	2	10560	-103.63	10.23	4	7	1	15840	-62.26	9.82
9	0	3	880	-109.28	10.18	4	6	2	83160	-66.15	10.05
8	4	0	34650	-91.45	8.75	4	5	3	110880	-72.56	9.69
8	3	1	110880	-92.22	9.69	4	4	4	34650	-81.48	8.75
8	2	2	83160	-95.50	10.05	3	8	1	880	-55.57	10.18
8	1	3	15840	-101.30	9.82	3	7	2	10560	-59.61	10.23
8	0	4	495	-109.61	9.09	3	6	3	24640	-66.17	9.79
7	5	0	44352	-83.33	8.73	3	5	4	12320	-75.24	8.80
7	4	1	221760	-84.25	9.67	2	8	2	396	-53.38	10.69
7	3	2	266112	-87.69	10.00	2	7	3	2112	-60.10	10.06
7	2	3	88704	-93.64	9.69	2	6	4	1848	-69.32	8.96
7	1	4	6336	-102.10	8.81	1	8	3	48	-54.34	10.65
6	6	0	25872	-75.53	8.73	1	7	4	96	-63.72	9.37
6	5	1	206976	-76.60	9.67	0	8	4	1	-58.43	10.17
6	4	2	388080	-80.19	9.99						

Using the identity

$$\binom{N}{m}^{-1}\binom{N-p}{m-t}$$
$$= \frac{m(m-1)\cdots(m-t+1)\,(N-m)(N-m-1)\cdots(N-m-p+t+1)}{N(N-1)\cdots(N-p+1)},$$

$$(14)$$

we obtain

$$\frac{m_r}{N_r}\frac{m_s}{N_s}\frac{N_t-m_t}{N_t}\frac{N_u-m_u}{N_u},$$

for the propagator of the average trace of $H^2(1,0)$ for $r \neq s \neq t \neq u$. In contrast, if all the contractions are in the same orbit, *i.e.*, $r = s = t = u$, we have $(p_r, t_r) = (4, 2)$ and the propagator is

$$\frac{m_r(m_r - 1)(N_r - m_r)(N_r - m_r - 1)}{N_r(N_r - 1)(N_r - 2)(N_r - 3)}.$$

From these two limiting cases it is not difficult to work out the propagator for the general case.

The expression for the average trace for the square of the unitary rank 2 part of the Hamiltonian, incorporating both O_t^p from eq. (13) and the propagator, can now be written in the form

$$\langle H^2(2,0)\rangle^{(m)} = \frac{1}{4}\sum_{rstu} \frac{(N_r - m_r)(N_s - m_s - \delta_{rs})m_t(m_u - \delta_{tu})}{(N_r - \delta_{rt} - \delta_{ru})(N_s - \delta_{st} - \delta_{su} - \delta_{rs})N_t(N_u - \delta_{tu})}$$

$$\times \sum_{\Gamma}[\Gamma]\{V_{rstu}^\Gamma\}^2(1 + \delta_{rs})(1 + \delta_{tu})$$

$$= \sum_{\substack{r \leq s \\ t \leq u}} \frac{(N_r - m_r)(N_s - m_s - \delta_{rs})m_t(m_u - \delta_{tu})}{(N_r - \delta_{rt} - \delta_{ru})(N_s - \delta_{st} - \delta_{su} - \delta_{rs})N_t(N_u - \delta_{tu})}$$

$$\times \sum_{\Gamma}[\Gamma]\{V_{rstu}^\Gamma\}^2. \tag{15}$$

Together with eqs. (12) we have the complete expression for the variance for a configuration density distribution. In Table VI-1, the centroids and widths of the 45 configurations in the $(ds)^{12}$-space are given as illustrations.

VI.3   Partial width

The expression in eq. (15) lends itself readily to the calculation of the partial width $\sigma_m(m')$ related to σ_m through the variances,

$$\sigma_m^2 = \sum_{m'} \sigma_m^2(m').$$

It provides an estimate of the spreading of the configuration strengths due to excitations to intermediate states in configuration m'. For example, the internal width, with $m' = m$, gives the contribution due to interaction among the states within m. On the other hand, the external widths ($m' \neq m$) come from intermediate states outside the configuration m. In general, if the external widths for all the irreducible representations of a group are small, it implies that the group is an approximate symmetry of the system.

The partial width $\sigma_m(m')$ also gives an indication of the importance of configuration m' for the states in m. If one is interested in the states lying primarily in m, one may be able to truncate the active space by eliminating

m' if $\sigma_m(m')$ is small (and far away in terms of centroid energy difference). The details concerning truncating the shell model space will be discussed in Chapter X.

In eq. (15), each of the terms in the summation over the orbital indices r, s, t, and u connects the configuration m to a particular m' given by

$$m' = m + t + u - r - s.$$

That is, m' differs from m in that a particle in orbit r and a particle in orbit s are replaced by a particle in orbit t and a particle in orbit u respectively. In general, there is more than one way to go from m to m'. For example, all the terms with $\{r, s\} = \{t, u\}$ contribute to the internal width $\sigma_m(m)$. Hence, for a space without radially degenerate orbits, the internal width can be obtained from eq. (15) and is given by

$$\sigma_m(m) = \sum_{r \leq s} \frac{(N_r - m_r)(N_s - m_s - \delta_{rs})m_r(m_s - \delta_{rs})}{(N_r - 1 - \delta_{rs})(N_s - 2\delta_{rs} - 1)N_r(N_s - \delta_{rs})} \sum_{\Gamma} [\Gamma]\{V_{rsrs}^{\Gamma}\}^2.$$

Similar expressions can be obtained for the external widths connecting configuration m with configurations $m' \neq m$.

VI.4   Configuration skewness

To obtain the skewness of the density distribution in a configuration m, we need $\langle (H - \langle H \rangle)^3 \rangle^{(m)}$. In terms of the $\nu = 1$ and 2 parts of the Hamiltonian, we have

$$
\begin{aligned}
H^3 =\ & H^3(1,0) \\
& + H^2(1,0)H(2,0) + H(1,0)H(2,0)H(1,0) + H(2,0)H^2(1,0) \\
& + H(1,0)H^2(2,0) + H(2,0)H(1,0)H(2,0) + H^2(2,0)H(1,0) \\
& + H^3(2,0),
\end{aligned}
\tag{16}
$$

identical to eq. (V-61) for the scalar case. However, unlike the scalar case, the configuration averages of the three terms in the second line, and the three terms in the third line of eq. (16) are in general no longer equal to each other. This is due to the fact that $H(1,0)$ and $H(2,0)$ are not necessarily number-conserving in each orbit. As a result, cyclic permutation of the operators gives, for example, the result that in general

$$\ll H^2(1,0)H(2,0) \gg^{(m)}\ =\ \ll H(1,0)H(2,0)H(1,0) \gg^{(m')}$$

$$\neq\ \ll H(1,0)H(2,0)H(1,0) \gg^{(m)}$$

On the other hand, the fully contracted quantities O_t^p are similar for all three terms in the second line of eq. (16), and for the three terms in the third line. For the first one, $H_1(1,0)H_2(1,0)H(2,0)$, we have

$$O_{2(rstu)}^4(123) = \tilde{\mathcal{E}}_{rt}(1)\,\tilde{\mathcal{E}}_{su}(2)\,\beta_{rtsu}^0\,\delta_{rt}^{\text{a.m.}}\,\delta_{su}^{\text{a.m.}}, \tag{17}$$

similar to eq. (V-65). The multipole coefficient, β_{rtsu}^Δ, in analogy with eqs. (V-68) and (D-24), is given in terms of the two-body matrix elements of the unitary rank $\nu = 2$ part of the Hamiltonian in configuration averaging,

$$\beta_{rtsu}^\Delta = \sum_\Gamma (-1)^{s+t+\Gamma+\Delta}[\Delta]^{\frac{1}{2}}[\Gamma]\begin{Bmatrix} r & s & \Gamma \\ u & t & \Delta \end{Bmatrix} V_{rstu}^\Gamma. \tag{18}$$

We can obtain the fully contracted quantity O_t^p for $H_1(1,0)H(2,0)H_2(1,0)$ from eq. (17) by permuting operator number 3, the $\nu = 2$ operator, with $H_2(1,0)$,

$$O_{2(rstu)}^4(132) = \tilde{\mathcal{E}}_{rt}(1)\,\beta_{rtsu}^0\,\tilde{\mathcal{E}}_{su}(2)\,\delta_{rt}^{\text{a.m.}}\,\delta_{su}^{\text{a.m.}}. \tag{19}$$

The values of O_t^p given by eqs. (17) and (19) are the same, but the orbits in which the two left-contractions and the two right-contractions take place are different in general. For eq. (17) we have, assuming that r, s, t, and u refer to different orbits, $(p_r, t_r) = (p_s, t_s) = (1,1)$ and $(p_t, t_t) = (p_u, t_u) = (1,0)$, since the single-particle operators appear in the order $A^r B^t\, A^t B^u\, A^t A^u B^r B^s$. On the other hand, for eq. (19), we have $(p_r, t_r) = (p_u, t_u) = (1,1)$ and $(p_s, t_s) = (p_t, t_t) = (1,0)$, since the order of the single-particle operators is now $A^r B^t\, A^t A^u B^r B^s\, A^s B^u$. Hence, the three terms in eq. (16) with two $H(1,0)$'s and one $H(2,0)$ have the same O_t^p values but different propagators. The same is true for the three terms consisting of the product of one $H(1,0)$ and two $H(2,0)$'s.

In order to shorten the discussion, we shall work out only the propagator for the product of the $\nu = 2$ parts of the three Hamiltonians, *i.e.*, the last term of eq. (16). This is the most important term in the expression since in the absence of radially degenerate orbits the Hamiltonian does not have any $\nu = 1$ part, and only the $\nu = 2$ term can contribute.

There are two diagrams associated with $\langle H^3(2,0)\rangle^{(m)}$, one with the total numbers of contractions in all the orbits $(p,t) = (6,2)$ and the other with $(p,t) = (6,3)$. These are shown in Figs. V-4f and g. Let us first deal with $(p,t) = (6,3)$. The form of O_t^p is

$$O_{3(rstuxy)}^6(123) = -\sum_\Delta \frac{(-1)^{r-t-\Delta}}{[\Delta]^{\frac{1}{2}}}\,\beta_{rtsu}^\Delta(1)\,\beta_{xysu}^\Delta(2)\,\beta_{xytr}^\Delta(3), \tag{20}$$

corresponding to the second term of eq. (V-68).

We shall now try to find the propagator associated with this term. By definition, the indices for the single-particle creation operators are given by the first and third subscripts of the multipole coefficient β^A_{rtsu}, as can be seen by comparing indices on the two sides of eq. (18). The single-particle creation operators in eq. (20) are therefore in orbits rs, uy, and xt. The single-particle creation and annihilation operators in $H^3(2,0)$ occur in the order $A^r A^s B^t B^u \; A^u A^y B^s B^x \; A^x A^t B^y B^r$. Since, from left to right, we have the creation operators in orbits r, s, and y appearing before the annihilation operators they contract with, these are the left-contractions. Similarly the right-contractions are in orbits t, u, and x. If all six orbits involved are distinct from each other, the propagator associated with these six orbits for the trace is simply

$$\binom{N_r - 1}{m_r - 1}\binom{N_s - 1}{m_s - 1}\binom{N_y - 1}{m_y - 1}\binom{N_t - 1}{m_t}\binom{N_u - 1}{m_u}\binom{N_x - 1}{m_x}.$$

In terms of an average trace, the complete propagator for the case of six distinct orbits is given by eq. (14) as

$$\frac{m_r}{N_r}\frac{m_s}{N_s}\frac{m_y}{N_y}\frac{N_t - m_t}{N_t}\frac{N_u - m_u}{N_u}\frac{N_x - m_x}{N_x}.$$

The general form, allowing for the possibility that some or all the orbits are the same, is

$$\frac{m_r(m_s - \delta_{rs})(m_y - \delta_{ry} - \delta_{sy})}{N_r(N_s - \delta_{rs})(N_y - \delta_{ry} - \delta_{sy})}\frac{(N_t - m_t)}{(N_t - \delta_{rt} - \delta_{st} - \delta_{yt})}$$

$$\times \frac{(N_u - m_u - \delta_{tu})}{(N_u - \delta_{tu} - \delta_{ru} - \delta_{su} - \delta_{yu})}\frac{(N_x - m_x - \delta_{tx} - \delta_{ux})}{(N_x - \delta_{tx} - \delta_{ux} - \delta_{rx} - \delta_{sx} - \delta_{yx})},$$

obtained in the same way as eq. (15).

For $(p, t) = (6, 2)$, the fully contracted quantity O^p_t takes the form

$$O^6_{2(rstuxy)}(123) = \sum_\Gamma [\Gamma] V^\Gamma_{rstu}(1)\, V^\Gamma_{tuxy}(2)\, V^\Gamma_{xyrs}(3). \tag{21}$$

The single-particle operators in $H^3(2,0)$ for this diagram are in the order $A^r A^s B^t B^u \; A^t A^u B^x B^y \; A^x A^y B^r B^s$. The two left-contractions are in orbits r and s, and the four right-contractions are in orbits t, u, x, and y. The propagator for the average trace is then

$$\frac{m_r(m_s - \delta_{rs})}{N_r(N_s - \delta_{rs})}\frac{(N_t - m_t)(N_u - m_u - \delta_{tu})}{(N_t - \delta_{rt} - \delta_{st})(N_u - \delta_{tu} - \delta_{ru} - \delta_{su})}$$

$$\times \frac{(N_x - m_x - \delta_{tx} - \delta_{ux})(N_y - m_y - \delta_{ty} - \delta_{uy} - \delta_{xy})}{(N_x - \delta_{tx} - \delta_{ux} - \delta_{rx} - \delta_{sx})(N_y - \delta_{ty} - \delta_{uy} - \delta_{xy} - \delta_{ry} - \delta_{sy})}.$$

On permuting the three $H(2,0)$ operators, the diagram $(p,t) = (6,2)$ transforms into one with $(p,t) = (6,4)$, as we have already seen in the scalar case. Except for the permutation, the form of O_t^p remains the same as that of eq. (21),

$$O^6_{4(rstuxy)}(321) = \sum_\Gamma [\Gamma]\, V^\Gamma_{xyrs}(3)\, V^\Gamma_{tuxy}(2)\, V^\Gamma_{rstu}(1). \tag{22}$$

The order of the single-particle operators of $H^3(2,0)$ corresponding to this term is now $A^x A^y B^r B^s\; A^t A^u B^x B^y\; A^r A^s B^t B^u$. There are four left-contractions, associated with orbits t, u, x, and y, and the two right-contractions are in orbits r and s. The propagator for the average trace of the operator is therefore

$$\frac{m_x(m_y - \delta_{xy})(m_t - \delta_{xt} - \delta_{yt})(m_u - \delta_{xu} - \delta_{yu} - \delta_{tu})}{N_x(N_y - \delta_{xy})(N_t - \delta_{xt} - \delta_{yt})(N_u - \delta_{xu} - \delta_{yu} - \delta_{tu})}$$

$$\times \frac{(N_r - m_r)(N_s - m_s - \delta_{rs})}{(N_r - \delta_{xr} - \delta_{yr} - \delta_{tr} - \delta_{ur})(N_s - \delta_{rs} - \delta_{xs} - \delta_{ys} - \delta_{ts} - \delta_{us})}.$$

The final result for the configuration average trace of $H^3(2,0)$ is then

$$\langle H_1(2,0)\, H_2(2,0)\, H_3(2,0)\rangle^{(m)}$$

$$= \sum_{\substack{rst \\ xyx}} \left\{ -\frac{m_r(m_s - \delta_{rs})(m_y - \delta_{ry} - \delta_{sy})}{N_r(N_s - \delta_{rs})(N_y - \delta_{ry} - \delta_{sy})} \frac{(N_t - m_t)}{(N_t - \delta_{rt} - \delta_{st} - \delta_{yt})} \right.$$

$$\times \frac{(N_u - m_u - \delta_{tu})}{(N_u - \delta_{tu} - \delta_{ru} - \delta_{su} - \delta_{yu})} \frac{(N_x - m_x - \delta_{tx} - \delta_{ux})}{(N_x - \delta_{tx} - \delta_{ux} - \delta_{rx} - \delta_{sx} - \delta_{yx})}$$

$$\times \sum_\Delta \frac{(-1)^{r-t-\Delta}}{[\Delta]^{\frac{1}{2}}}\, \beta^\Delta_{rtsu}(1)\, \beta^\Delta_{xysu}(2)\, \beta^\Delta_{xytr}(3)$$

$$+ \left[\frac{m_r(m_s - \delta_{rs})}{N_r(N_s - \delta_{rs})} \frac{(N_t - m_t)(N_u - m_u - \delta_{tu})}{(N_t - \delta_{rt} - \delta_{st})(N_u - \delta_{tu} - \delta_{ru} - \delta_{su})} \right.$$

$$\times \frac{(N_x - m_x - \delta_{tx} - \delta_{ux})(N_y - m_y - \delta_{ty} - \delta_{uy} - \delta_{xy})}{(N_x - \delta_{tx} - \delta_{ux} - \delta_{rx} - \delta_{sx})(N_y - \delta_{ty} - \delta_{uy} - \delta_{xy} - \delta_{ry} - \delta_{sy})}$$

$$+ \frac{m_x(m_y - \delta_{xy})(m_t - \delta_{xt} - \delta_{yt})(m_u - \delta_{xu} - \delta_{yu} - \delta_{tu})}{N_x(N_y - \delta_{xy})(N_t - \delta_{xt} - \delta_{yt})(N_u - \delta_{xu} - \delta_{yu} - \delta_{tu})}$$

$$\times \left. \frac{(N_r - m_r)(N_s - m_s - \delta_{rs})}{(N_r - \delta_{xr} - \delta_{yr} - \delta_{tr} - \delta_{ur})(N_s - \delta_{rs} - \delta_{xs} - \delta_{ys} - \delta_{ts} - \delta_{us})} \right]$$

$$\times \left. \sum_\Gamma [\Gamma] V^\Gamma_{rstu}(1)\, V^\Gamma_{tuxy}(2)\, V^\Gamma_{xyrs}(3) \right\}, \tag{23}$$

where we have allowed the possibility for the three Hamiltonians to be different.

We shall not attempt to work out the trace of any of the $H^4(2,0)$ terms here. This is not difficult to do since the forms of O_t^p, except for the summation over orbital indices, are already given by eqs. (V-69) for the scalar case. The only additional work for the configuration case here is to construct the propagator, which as we have seen in the examples given above depends on the orbits in which the contractions take place. The form of the propagator, while complex in appearance and tedious to write down, is actually quite simple to be expressed in the form of an algorithm for a computer program. In fact, all three steps, (1) finding the basic diagrams, (2) constructing the form of O_t^p, and (3) writing a computer program to evaluate the O_t^p and the propagators can be carried out using a computer (Chang and Wong, 1979, 1980, Chang, Draayer and Wong, 1982). This is made possible primarily because of the basic simplicity of the method used and, as a result, the tedium of the algebraic work can hence be delegated to a machine.

Chapter VII

Isospin Projection

The space of m particles may be partitioned according to the isospin T. The emphasis here is somewhat different than for the configuration subdivision described in the previous chapter. Since isospin can be assumed to be a good quantum number for most purposes, a separate distribution for each isospin is of physical interest in itself. Furthermore, it is known that in even-even nuclei the low-lying states for different isospins are well separated in energy, with the lowest ones having the minimum isospin; different isospin distributions are expected to have different centroid energies. Hence, a significant part of the spread in strength distributions in the complete m-particle space may be taken up by the separations between individual isospin distributions, and fixed-isospin distributions are expected to be narrower on the average than for the complete distribution in the m-particle space.

The simplest method to obtain moments in isospin subspace is by projecting the fixed-isospin traces from the traces in the m-particle space for scalar-isospin averaging, and from the traces in the m-configuration subspace for configuration-isospin averaging. The projection method is applicable here since a particle can take on only the isospin value $t = 1/2$. In contrast, projection of spin is much harder, since the single-particle spin j can take on a wide range of values.

VII.1 Dimension of a space with fixed isospin

In a subspace with proton number m_p and neutron number m_n, the projection Z of the isospin on the quantization axis, by definition, is equal to $(m_p - m_n)/2$. The possible isospin in the subspace therefore ranges from $|m_p - m_n|/2$ to $(m_p + m_n)/2$, and the total number of states, d_Z, in the subspace is given by the sum

$$d_Z = \sum_{T \geq Z} d_{T,Z}, \qquad (1)$$

where $d_{T,Z}$ is the number of states in the subspace with isospin T and projection Z.

In contrast, in the space of m particles, with all possible partitions into protons and neutrons, the total number of states with a given isospin T is

$$d_T = \sum_{Z=-T}^{T} d_{T,Z} = (2T + 1)d_{T,Z}, \tag{2}$$

since $d_{T,Z}$ is the same for all the Z-values of a given T. For the same reason we obtain from eq. (1),

$$d_{Z+1} = \sum_{T \geq Z+1} d_{T,Z+1} = \sum_{T \geq Z+1} d_{T,Z} = d_Z - d_{T=Z,Z}. \tag{3}$$

As a result, we have

$$d_T = (2T + 1)d_{T,Z} = (2T + 1)\{d_{Z=T} - d_{Z=T+1}\}, \tag{4}$$

using eqs. (2) and (3).

For a space made up of a total of N single-particle states divided equally between protons and neutrons, we have

$$d_Z = \binom{\frac{N}{2}}{m_p}\binom{\frac{N}{2}}{m_n} = \binom{\frac{N}{2}}{\frac{m}{2} + Z}\binom{\frac{N}{2}}{\frac{m}{2} - Z}, \tag{5}$$

since $m_p = m/2 + Z$ and $m_n = m/2 - Z$. Similarly,

$$d_{Z+1} = \binom{\frac{N}{2}}{\frac{m}{2} + Z + 1}\binom{\frac{N}{2}}{\frac{m}{2} - Z - 1}.$$

The difference,

$$d_Z - d_{Z+1} = \binom{\frac{N}{2}}{\frac{m}{2} + Z}\binom{\frac{N}{2}}{\frac{m}{2} - Z} - \binom{\frac{N}{2}}{\frac{m}{2} + Z + 1}\binom{\frac{N}{2}}{\frac{m}{2} - Z - 1}$$

$$= \binom{\frac{N}{2} + 1}{\frac{m}{2} + Z + 1}\binom{\frac{N}{2} + 1}{\frac{m}{2} - Z}$$

$$\times \left\{ \frac{(\frac{m}{2} + Z + 1)(\frac{N}{2} - \frac{m}{2} + Z + 1)}{(\frac{N}{2} + 1)^2} - \frac{(\frac{N}{2} - \frac{m}{2} - Z)(\frac{m}{2} - Z)}{(\frac{N}{2} + 1)^2} \right\}$$

$$= \binom{\frac{N}{2} + 1}{\frac{m}{2} + Z + 1}\binom{\frac{N}{2} + 1}{\frac{m}{2} - Z}\left\{ \frac{ZN + 2Z + \frac{N}{2} + 1}{(\frac{N}{2} + 1)^2} \right\}$$

$$= \frac{2(2Z + 1)}{N + 2}\binom{\frac{N}{2} + 1}{\frac{m}{2} + Z + 1}\binom{\frac{N}{2} + 1}{\frac{m}{2} - Z}, \tag{6}$$

is the number of states $d_{T,Z}$ for $T = Z$. Using eq. (2), we obtain for the number of states with a given T

$$d_T = \frac{2(2T + 1)^2}{N + 2}\binom{\frac{N}{2} + 1}{\frac{m}{2} + T + 1}\binom{\frac{N}{2} + 1}{\frac{m}{2} - T}, \tag{7}$$

a well known result.

VII.2   Dimension of configuration-isospin space

We can also write down in a similar way $d_{m,T}$, the number of states in configuration m with isospin T. Let us denote a proton-neutron configuration by $m_{pn} = (m_{1p}, m_{1n}, m_{2p}, m_{2n}, \ldots)$ where m_{rp} and m_{rn} are, respectively, the number of protons and neutrons in orbit r, and $m_r = m_{rp} + m_{rn}$. The total number of protons in this configuration is then $m_p = \sum_r m_{rp}$, and the total number of neutrons $m_n = \sum_r m_{rn}$. In general, there is more than one m_{pn} configuration in the subspace of configuration m with the value $Z = (m_p - m_n)/2$. Using the same argument as in arriving at eq. (IV-2), the dimension of a proton-neutron configuration is found to be

$$d_{m_{pn},Z} = \prod_r \binom{\frac{N_r}{2}}{m_{rp}} \binom{\frac{N_r}{2}}{m_{rn}}.$$

From this we can obtain $d_{m,Z}$ by summing over all the m_{pn}-subspaces in configuration m with the given Z value.

A proton-neutron configuration with isospin projection $(Z + 1)$ can be generated from a configuration with isospin projection Z by changing a neutron to a proton in any one of the orbits allowed by the Pauli principle. Using the same arguments as those in arriving at eq. (4), we obtain

$$d_{m,T} = (2T + 1)(d_{m,Z} - d_{m,Z+1}). \tag{8}$$

It is not possible to write eq. (8) in the form of eq. (7) since in each orbit there is no longer the simple relation between (m_p, m_n) and (m, Z) as given by eq. (5) for scalar averaging.

VII.3   Centroids of scalar-isospin distributions

The centroid of a density distribution for a given mT is the average trace of the Hamiltonian in the subspace of m particles with isospin T,

$$\overline{E}_{m,T} = \langle H \rangle^{(m,T)} = \frac{1}{d_{m,T}} \sum_x \langle mTx|H|mTx \rangle, \tag{9}$$

where x stands for all the labels of a state in the space other than m and T. For the one-body Hamiltonian given by eq. (IV-6),

$$H(1) = \sum_r \epsilon_r \hat{n}_r,$$

all the different isospin subspaces have the same centroid since there are no terms, such as the Coulomb energy, that depend explicitly on the difference between protons and neutrons.

For later convenience, it is useful to examine this simple point in a different way. Since a single-particle creation or annihilation operator has $t = 1/2$, there is only one isoscalar one-body operator that can be formed from the coupling of a creation operator with an annihilation operator, and the number operator $\hat{n}$ is such an operator. As a result, all the one-body, isoscalar operators in the space must be proportional to $\hat{n}$. Since the expectation value of $\hat{n}$ depends only on the number of active nucleons, it cannot distinguish between different isospin subspaces. The average of $H(1)$ is therefore independent of T,

$$\langle H(1) \rangle^{m,T} = m\bar{\epsilon}, \tag{10}$$

where $\bar{\epsilon}$ is the average single-particle energy defined in eq. (IV-6). The result is the same as that in the m-particle space.

On the other hand, there are two two-body, isoscalar operators, and they can be distinguished by whether the intermediate isospin rank is coupled to 0 or 1. As a result, a splitting of the centroid energies of different isospin distributions can emerge for $m \geq 2$. We can take these two operators to be $\binom{\hat{n}}{2}$ and $(T^2 - \frac{3}{4}\hat{n})$. The particular forms of these operators ensure that they are purely particle rank-two operators. The part of $H(2)$ that contributes to the (m, T) trace, or the *trace-equivalent* operator of $H(2)$, must be a function of these two operators alone and can be written in the form

$$H_{\mathrm{tr}}(2) = \alpha \binom{\hat{n}}{2} + \beta \left(T^2 - \tfrac{3}{4}\hat{n} \right). \tag{11}$$

We can find the two constants, α and β, in this expression from the average values of $H(2)$ in the two-particle space. For the $T = 0$ subspace, we have

$$\langle H(2) \rangle^{(m=2,T=0)} \equiv \overline{W^0} = \frac{1}{\sum_J (2J+1)} \sum_{r \leq s; J} (2J+1) W^{JT=0}_{rsrs}, \tag{12}$$

and for the $T = 1$ subspace,

$$\langle H(2) \rangle^{(m=2,T=1)} \equiv \overline{W^1} = \frac{1}{\sum_J (2J+1)} \sum_{r \leq s; J} (2J+1) W^{JT=1}_{rsrs}, \tag{13}$$

where W^{JT}_{rstu} is the two-body matrix element defined in eq. (D-21). Comparing these results with the trace of $H_{\mathrm{tr}}(2)$ given by eq. (11), we have

$$\alpha = \frac{1}{4} \left(\overline{W^0} + 3\overline{W^1} \right), \quad \text{and} \quad \beta = \frac{1}{2} \left(\overline{W^1} - \overline{W^0} \right).$$

Instead of α and β, we can now use $\overline{W^0}$ and $\overline{W^1}$ as the defining quantities of $H_{\mathrm{tr}}(2)$.

We can now combine the results of eqs. (10), (12), and (13), and find an expression of the trace-equivalent operator for the complete Hamiltonian. From the mT-centroids for two particles,

$$\overline{E}_{2,0} = 2\overline{\epsilon} + \overline{W^0}, \quad \text{and} \quad \overline{E}_{2,1} = 2\overline{\epsilon} + \overline{W^1},$$

we obtain

$$H_{\text{tr}} = \overline{\epsilon}\hat{n} + \frac{1}{4}\left(\overline{W^0} + 3\overline{W^1}\right)\binom{\hat{n}}{2} + \frac{1}{2}\left(\overline{W^1} - \overline{W^0}\right)\left(T^2 - \frac{3}{4}\hat{n}\right). \qquad (14)$$

We can regard $\overline{\epsilon}$, $\overline{W^0}$, and $\overline{W^1}$ as the defining averages of the operator H_{tr}.

From eq. (14), we obtain the result that the centroid difference between two adjacent T-distributions for a given particle number m is

$$\overline{E}_{m,T} - \overline{E}_{m,T-1} = \left(\overline{W^1} - \overline{W^0}\right)T, \qquad (15)$$

and the difference between the maximum T ($= m/2$ for $m \leq N/2$ and $(N - m)/2$ for $m > N/2$) and minimum T ($= 0$ for even m and $1/2$ for odd m) is

$$\overline{E}_{m,T_{\max}} - \overline{E}_{m,T_{\min}} = \begin{cases} \frac{1}{8}m(m + 2)\left(\overline{W^1} - \overline{W^0}\right) & \text{for } m \text{ even,} \\ \frac{1}{8}(m - 1)(m + 3)\left(\overline{W^1} - \overline{W^0}\right) & \text{for } m \text{ odd.} \end{cases}$$
$$(16)$$

The centroid energy differences given by eqs. (15) and (16) provide some guidance regarding the isospin splitting of nuclear states. Note that these quantities are only functions of the difference between $\overline{W^1}$ and $\overline{W^0}$.

VII.4   Widths of scalar-isospin distributions

For the width of an isospin distribution we need to find $(H^2)_{\text{tr}}$, the trace equivalent operator of H^2. It is an operator with maximum particle rank 4, and there is a total of eight independent operators up to particle rank 4 that can be constructed in the mT-subspace. For convenience we shall express them as:

$$\begin{array}{cccc} \hat{n}, & \hat{n}^2, & \hat{n}^3, & \hat{n}^4, \\ & T^2, & \hat{n}T^2, & \hat{n}^2T^2, \\ & & & T^4. \end{array}$$

There are also exactly eight (m, T)-subspaces for $m \leq 4$:

$$(1, \tfrac{1}{2}), \quad (2, 0), \quad (2, 1), \quad (3, \tfrac{1}{2}), \quad (3, \tfrac{3}{2}), \quad (4, 0), \quad (4, 1), \quad \text{and} \quad (4, 2).$$

In principle, we can parametrize $(H^2)_{\text{tr}}$ in terms of these eight operators in the form

$$\begin{aligned} (H^2)_{\text{tr}} = {} & \alpha_1\hat{n} + \alpha_2\hat{n}^2 + \alpha_3\hat{n}^3 + \alpha_4\hat{n}^4 \\ & + \alpha_5T^2 + \alpha_6\hat{n}T^2 + \alpha_7\hat{n}^2T^2 + \alpha_8T^4. \end{aligned} \qquad (17)$$

Table VII-1. Dimensions, centroids, and widths of mT-configurations for CW particle interaction in $(ds)^6$-space

$d_{5/2}d_{3/2}s_{1/2}$			$T = 0$			$T = 1$			$T = 2$			$T = 3$		
			d	C	σ	d	C	σ	d	C	σ	d	C	σ
6	0	0	175	-45.15	7.20	189	-42.94	6.84	35	-38.51	6.18	1	-31.86	5.40
5	1	0	840	-39.37	7.16	1176	-36.80	6.54	360	-31.89	5.39	24	-24.82	3.88
5	0	1	420	-41.26	7.50	588	-39.08	7.08	180	-34.70	6.30	12	-28.09	5.32
4	2	0	1680	-32.99	7.03	2400	-30.58	6.43	870	-25.59	5.15	90	-18.11	3.05
4	1	1	1680	-35.61	7.36	2640	-33.28	6.85	1080	-28.57	5.75	120	-21.57	4.08
4	0	2	420	-39.57	7.15	540	-37.56	6.74	165	-33.55	5.78	15	-27.32	4.15
3	3	0	1480	-26.68	7.06	2160	-24.14	6.44	760	-19.24	5.20	80	-11.73	3.14
3	2	1	2480	-29.60	7.28	3960	-27.30	6.75	1720	-22.61	5.63	240	-15.38	3.64
3	1	2	1200	-34.32	7.10	1800	-32.01	6.52	680	-27.67	5.37	80	-21.31	3.14
3	0	3	140	-40.87	6.62	180	-38.69	5.86	40	-34.31	4.53			
2	4	0	645	-20.05	7.19	855	-17.67	6.64	261	-12.67	5.47	15	-5.68	4.15
2	3	1	1560	-23.48	7.34	2448	-21.19	6.84	1008	-16.53	5.76	120	-9.51	4.08
2	2	2	1266	-28.52	7.03	1854	-26.36	6.50	726	-21.96	5.33	90	-15.64	3.05
2	1	3	288	-35.78	6.53	408	-33.48	5.79	120	-28.74	4.01			
2	0	4	21	-44.36	5.26	15	-42.15	4.98						
1	5	0	120	-13.57	7.57	144	-10.94	6.97	24	-6.33	6.23			
1	4	1	420	-17.21	7.56	612	-14.96	7.09	204	-10.44	6.20	12	-3.98	5.32
1	3	2	504	-22.75	7.17	720	-20.48	6.63	240	-16.40	5.62	24	-10.29	3.88
1	2	3	192	-30.32	6.56	264	-28.07	5.81	72	-23.50	4.19			
1	1	4	24	-39.86	5.48	24	-37.01	4.79						
0	6	0	10	-6.44	7.83	6	-4.46	7.79						
0	5	1	40	-10.68	7.82	48	-8.68	7.58	8	-4.71	7.27			
0	4	2	75	-16.42	7.25	81	-14.67	7.02	19	-11.08	6.31	1	-5.27	5.40
0	3	3	40	-24.55	6.68	48	-22.56	5.97	8	-18.59	5.03			
0	2	4	10	-34.19	5.30	6	-32.21	5.13						

and solve for the unknown coefficients α_i using as input the eight average traces $\langle H^2 \rangle^{m,T}$ for $m \leq 4$. Such a set of input is sometimes referred to as the *elementary net*, since it is the set that comes to mind most readily. However, it is not the most economical set to use since, as input, we need to calculate all the averages of H^2 up to $m = 4$.

As far as eq. (17) is concerned, we can determine the unknown coefficients α_i using the average traces in eight arbitrary (m, T)-subspaces, not necessarily those in the elementary net with $m \leq 4$. To minimize the effort involved in constructing the input, it is convenient to make use of traces over subspaces with m near N. In particular we can take the $(m, T) = (N, 0)$ and $(N - 1, 1/2)$ subspaces. Since these subspaces have only one state in each of them, they have isospin distributions with zero width, and consequently $\langle H^2 \rangle^{(m,T)} = (\langle H \rangle^{(m,T)})^2$. Similarly, the $(m, T) = (N/2, N/4)$ subspace contains only a single state formed by $N/2$ identical particles filling all the available proton (or neutron) single-particle states. As a result, its distribution in isospin space is again a delta function with zero width.

The amount of effort required to prepare the input of $(H^2)_{tr}$ can also be reduced by making use of particle-hole transformations. In general, the widths for isospin subspaces of m particles can be shown to be the same as those in the corresponding subspaces for $(N-m)$ particles. As we have seen earlier, the one-body term in the Hamiltonian contributes only to the centroid and not to the width of a fixed isospin subspace. The two-body operator, $((A^t \times A^u)^\Delta \times (B^r \times B^s)^\Delta)^0$, changes into $((B^r \times B^s)^\Delta \times (A^t \times A^u)^\Delta)^0$ under a particle-hole transformation. These two operators are also related through the commutation relations (C-13) between single-particle operators and are given by

$$((B^r \times B^s)^\Delta \times (A^t \times A^u)^\Delta)^0 = -[\Delta]^{\frac{1}{2}}\{\delta_{rt}\delta_{su} - (-1)^{r+s+\Delta}\delta_{ru}\delta_{st}\}$$

$$+ \left[\frac{\Delta}{s}\right]^{\frac{1}{2}}\{\delta_{rt}(A^u \times B^s)^0 - (-1)^{t+u+\Delta}\delta_{ru}(A^t \times B^s)^0\}$$

$$- (-1)^{r+s+\Delta}\left[\frac{\Delta}{r}\right]^{\frac{1}{2}}\{\delta_{st}(A^u \times B^r)^0 - (-1)^{t+u+\Delta}\delta_{su}(A^t \times B^r)^0\}$$

$$+ ((A^t \times A^u)^\Delta \times (B^r \times B^s)^\Delta)^0. \tag{18}$$

For a scalar two-body operator, a matrix element in the $(N - m)$-particle space is related to its compliment in the m-particle space through a particle-hole transformation,

$$\langle (N-m)\Gamma x'_c|((A^t \times A^u)^\Delta \times (B^r \times B^s)^\Delta)^0|(N-m)\Gamma x_c\rangle$$

$$= \langle m\Gamma x'|((B^t \times B^u)^\Delta \times (A^r \times A^s)^\Delta)^0|m\Gamma x\rangle, \tag{19}$$

where the states $|(N - m)\Gamma x_c\rangle$ and $|m\Gamma x\rangle$ are compliments of each other as defined in eq. (V-41). On substituting eq. (18) into the right-hand side of eq. (19), we find that the matrix elements of $H(2)$ in the $(N - m)$-space are equal to their compliments in the m-particle space, except for the zero- and one-body terms in eq. (18). Since these two terms do not contribute to the variance of the isospin distribution, we obtain the result

$$\sigma^2_{m,T} = \sigma^2_{(N-m),T}. \tag{20}$$

Eq. (20) will not be correct in configuration averaging if there are radially degenerate orbits. In such cases, we have both the one-body terms in eq. (18) and those in $H(1)$ contributing also to the variance.

Instead of writing $(H^2)_{tr}$ in terms of the elementary net, we can now write it in terms of the average traces in the following eight (m,T)-subspaces: $(1,0)$, $(2,0)$, $(2,1)$, $(N,0)$, $(N-1,1/2)$ $(N-2,0)$, $(N-2,1)$, and $(N/2, N/4)$. For four of these eight subspaces, $(1,0)$, $(N,0)$, $(N - 1,1/2)$, and $(N/2, N/4)$, the variances vanish. With the help of the relation given in eq. (20) for $m = 2$, the variance of an (m, T)-distribution can be simplified to an expression involving only $\sigma_{2,0}^2$ and $\sigma_{2,1}^2$ as the input,

$$
\sigma^2(m, T) = \frac{m(m + 2) - 4T(T + 1)}{8N(N - 2)} \{(N - m)(N - m + 2) - 4T(T + 1)\}\sigma_{2,0}^2
$$

$$
+ \frac{1}{8(N + 2)N(N - 4)(N - 6)} \Big\{ 16(3N^2 - 14N + 24)T^2(T + 1)^2
$$

$$
+ [4(N + 12)(N + 2)N(N - 2) - 8(N + 4)(5N - 6)m(N - m)]T(T + 1)
$$

$$
+ 3(N + 4)(N + 2)m(m - 2)(N - m)(N - m - 2)\Big\}\sigma_{2,1}^2. \tag{21}
$$

This expression was first given by French (1969).

In Table VII-1, the calculated values of the centroid, $\overline{E}_{m,T}$, and width, $\sigma_{m,T}$, are given for the Chung-Wildenthal (1979) particle interactions in the $(ds)^6$-space to provide a feeling for the magnitudes of the quantities discussed so far.

VII.5 Isoscalar operators

For more general operators the method used above in averaging H and H^2 may not be applicable. However, the simplicity of the isospin subspace allows us to project out the fixed-T traces, $\ll \hat{O} \gg^{(T,Z)}$, of an operator $\hat{O}$ from $\ll \hat{O} \gg^{(Z)}$, the trace calculated in the proton-neutron subspace. The method is essentially the same as the one used to obtain the fixed-T dimensions in section 1. By definition, we have for traces of an arbitrary operator $\hat{O}$,

$$
\ll \hat{O} \gg^{(Z)} = \sum_{T \geq Z} \ll \hat{O} \gg^{(T,Z)} . \tag{22}
$$

Hence, with a set of traces of an operator $\hat{O}$ for different Z values, we can extract the values of fixed-T traces of $\hat{O}$ in a way similar to the procedure used in arriving at eqs. (4) and (8). However, before we can take the difference between two traces, it is necessary to display explicitly the Z-dependence of the expectation value, and hence the trace, of $\hat{O}$ in a state of given T.

Let us illustrate the point by the simple example of an isoscalar operator $\hat{O}^0$. Since in this case the expectation value is independent of Z, we obtain from eq. (22)

$$
\ll \hat{O}^0 \gg^{(Z)} - \ll \hat{O}^0 \gg^{(Z+1)} = \ll \hat{O}^0 \gg^{(T,Z)} .
$$

Hence the average trace is given by

$$\langle \hat{O}^0 \rangle^{(T=Z)} = \frac{1}{d_{T,Z}} \left\{ \ll \hat{O}^0 \gg^{(Z)} - \ll \hat{O}^0 \gg^{(Z+1)} \right\}$$

$$= \frac{2T+1}{d_T} \left\{ \ll \hat{O}^0 \gg^{(Z)} - \ll \hat{O}^0 \gg^{(Z+1)} \right\}, \qquad (23)$$

where the final result is obtained using eq. (2). For an operator with nonzero isospin rank, the situation is only slightly more involved, as we shall see in the next section.

VII.6   Operators of arbitrary isospin rank

For an operator $\hat{O}^\omega$ with isospin rank ω, the Z-dependence of the expectation value can be separated out using the Wigner-Eckart theorem given in eq. (C-3),

$$\langle TZ\alpha | \hat{O}^\omega | TZ\alpha \rangle = (-1)^{T-Z} \begin{pmatrix} T & \omega & T \\ Z & 0 & Z \end{pmatrix} \langle T\alpha \| \hat{O}^\omega \| T\alpha \rangle, \qquad (24)$$

where α represents all the other quantum numbers required to specify the state. On the right hand side of eq. (24), the Z-dependence of the expectation value is completely contained in the isospin $3j$-coefficient.

For $Z_{\max}$, the maximum Z value in the space, there is only one possible T value, $T_{\max} = Z_{\max}$. Hence,

$$\ll \hat{O}^\omega \gg^{(T_{\max}, Z_{\max})} = \ll \hat{O}^\omega \gg^{(Z_{\max})} . \qquad (25)$$

For the trace in the $Z = (Z_{\max} - 1)$ space,

$$\ll \hat{O}^\omega \gg^{(Z_{\max}-1)} = \ll \hat{O}^\omega \gg^{(T_{\max}, Z_{\max}-1)} + \ll \hat{O}^\omega \gg^{(T_{\max}-1, Z_{\max}-1)} . \qquad (26)$$

Using eq. (24), we can obtain $\ll \hat{O}^\omega \gg^{(T_{\max}, Z_{\max}-1)}$ from eq. (26) from the fact that

$$\frac{(-1)^{T-Z+1} \ll \hat{O}^\omega \gg^{(T,Z-1)}}{\begin{pmatrix} T & \omega & T \\ Z-1 & 0 & Z-1 \end{pmatrix}} = \sum_\alpha \langle T\alpha \| \hat{O}^\omega \| T\alpha \rangle = \frac{(-1)^{T-Z} \ll \hat{O}^\omega \gg^{(T,Z)}}{\begin{pmatrix} T & \omega & T \\ Z & 0 & Z \end{pmatrix}}.$$

Since the number of states in the (T, Z)-subspace is equal to that in the $(T, Z - 1)$-subspace, we obtain

$$\ll \hat{O}^\omega \gg^{(T_{\max}-1, Z_{\max}-1)}$$

$$= \ll \hat{O}^\omega \gg^{(Z_{\max}-1)} - \frac{(-1) \begin{pmatrix} T_{\max} & \omega & T_{\max} \\ Z_{\max}-1 & 0 & Z_{\max}-1 \end{pmatrix}}{\begin{pmatrix} T_{\max} & \omega & T_{\max} \\ Z_{\max} & 0 & Z_{\max} \end{pmatrix}} \ll \hat{O}^\omega \gg^{(T_{\max}, Z_{\max})},$$

where the value of $\ll \hat{O}^\omega \gg^{(T_{\max}, Z_{\max})}$ is obtained from eq. (25). Continuing the same process for successively lower T values, we obtain the general expression

$$\ll \hat{O}^\omega \gg^{(T, Z=T)} = \ll \hat{O}^\omega \gg^{(Z)}$$

$$- \sum_{T'=T+1}^{T_{\max}} \frac{(-1)^{T'-T} \left(\begin{smallmatrix} T' & \omega & T' \\ T & 0 & T \end{smallmatrix} \right)}{\left(\begin{smallmatrix} T' & \omega & T' \\ T' & 0 & T' \end{smallmatrix} \right)} \ll \hat{O}^\omega \gg^{(T', Z=T')} . \qquad (27)$$

In this way the traces of operator $\hat{O}$ for successively lower T values are obtained from the traces of fixed-Z traces for $Z = T$ and the fixed-T traces for the higher T values. Since the trace of $\hat{O}^\omega$ in subspaces of the same T but different Z are related through the Wigner-Eckart theorem, the results of eq. (27) can be used to obtain the fixed-(T, Z) traces of other Z values. For isoscalar operators, $\omega = 0$; since

$$(-1)^{T'-T} \left(\begin{matrix} T' & \omega & T' \\ T' & 0 & T' \end{matrix} \right)^{-1} \left(\begin{matrix} T' & \omega & T' \\ T & 0 & T \end{matrix} \right) = 1,$$

eq. (27) becomes identical to eq. (22).

VII.7   Isospin rank of operators in proton-neutron space

The starting point of an isospin trace calculation is the trace in proton-neutron space. To calculate a trace treating protons and neutrons as distinguishable particles, it is necessary first to express the operator in terms of proton and neutron single-particle creation and annihilation operators. In such cases, the operator no longer has a definite isospin rank. Let us illustrate the point using single-particle operators and ignoring all reference to spin for simplicity. The isospin convention used here is that a proton has isospin $t = 1/2$ and z-projection $+1/2$. A neutron, in turn, has $(t, z) = (1/2, -1/2)$. In the proton-neutron space, let $A(p)$ and $B(p)$ be the single-proton creation and annihilation operators, respectively. Similarly for neutrons we shall use $A(n)$ and $B(n)$. No isospin ranks enter here since we are considering protons and neutrons as distinguishable particles. To make the connection to isospin we need to find the relations between these two types of single-particle operators.

In the isospin formalism, $A_{1/2}^{1/2}$ creates a proton single-particle state when acting on the vacuum. Similarly, $A_{-1/2}^{1/2}$ creates a single-neutron state. Hence, we establish the relations

$$A_{1/2}^{1/2} = A(p), \qquad A_{-1/2}^{1/2} = A(n). \qquad (28a)$$

Eq. (28a) can be regarded as the definition of $A(p)$ and $A(n)$. Once these relations are fixed, the isospin ranks of annihilation operators follow from them and are given by $(t, z) = (1/2, -1/2)$ for a proton and $(1/2, 1/2)$ for a neutron. No freedom is left in defining the corresponding relations between the annihilation operators since they are the adjoints of the creation operators. Because of the relation given in eq. (C-5a), we have in isospin formalism,

$$\langle t = \tfrac{1}{2}, z = \tfrac{1}{2} | A_{1/2}^{1/2} | 0 \rangle = (-1)^{\frac{1}{2} - \frac{1}{2}} \langle 0 | \overline{A}_{-1/2}^{1/2} | t = \tfrac{1}{2}, z = \tfrac{1}{2} \rangle$$

$$= + \langle 0 | B_{-1/2}^{1/2} | t = \tfrac{1}{2}, z = \tfrac{1}{2} \rangle, \qquad (29a)$$

and

$$\langle t = \tfrac{1}{2}, z = -\tfrac{1}{2} | A_{-1/2}^{1/2} | 0 \rangle = (-1)^{\frac{1}{2} - (-\frac{1}{2})} \langle 0 | \overline{A}_{1/2}^{1/2} | t = \tfrac{1}{2}, z = -\tfrac{1}{2} \rangle$$

$$= - \langle 0 | B_{1/2}^{1/2} | t = \tfrac{1}{2}, z = -\tfrac{1}{2} \rangle, \qquad (29b)$$

where the bar indicates the adjoint of an operator. On the other hand, in the proton-neutron space,

$$\langle p | A(\mathrm{p}) | 0 \rangle = \langle 0 | B(\mathrm{p}) | p \rangle, \qquad (30a)$$

and

$$\langle n | A(\mathrm{n}) | 0 \rangle = \langle 0 | B(\mathrm{n}) | n \rangle, \qquad (30b)$$

without the phase factor due to isospin. Comparing eqs. (30) with (29), we have

$$B_{-1/2}^{1/2} = + B(\mathrm{p}), \qquad B_{1/2}^{1/2} = - B(\mathrm{n}), \qquad (28b)$$

for the relations between annihilation operators.

It is perhaps useful to give an example showing that the minus sign in eq. (28b) for a neutron is necessary. For this purpose, let us write down the number operator in isospin space, again ignoring all reference to spin. Since this operator counts particles without differentiating between neutrons and protons, it is necessarily an isoscalar operator. From eq. (D-3) we have

$$\hat{n} = \sqrt{2} \, (A^{1/2} \times B^{1/2})^0 = \sqrt{2} \sum_{z=-1/2}^{1/2} (-1)^{\frac{1}{2} + z} \sqrt{2} \begin{pmatrix} \tfrac{1}{2} & 0 & \tfrac{1}{2} \\ -z & 0 & z \end{pmatrix} A_{-z}^{1/2} B_z^{1/2},$$

where the statistical weight of $\sqrt{2}$ comes from isospin considerations alone. Putting in explicitly the values of the $3j$-coefficients for isospin, we obtain

$$\hat{n} = \sqrt{2} \left\{ \sqrt{\tfrac{1}{2}} A_{1/2}^{1/2} B_{1/2}^{1/2} - \sqrt{\tfrac{1}{2}} A_{1/2}^{1/2} B_{1/2}^{1/2} \right\}$$

$$= (A(\mathrm{p}) \times B(\mathrm{p}))^0 + (A(\mathrm{n}) \times B(\mathrm{n}))^0 = \hat{n}(\text{proton}) + \hat{n}(\text{neutron}). \quad (31)$$

We see that in this example the *extra* minus sign in eq. (28b) is required to cancel the opposite sign coming from the isospin coupling coefficient.

With the help of eqs. (28), we can transcribe any operator written in terms of proton-neutron single-particle creation and annihilation operators into isospin single-particle operators. In order to project out the part with a given isospin rank, we need in general to take a linear combination of terms each with different numbers of proton and neutron single-particle operators. This can best be illustrated with one-body operators.

An isoscalar, one-body operator can be written in the form

$$(A^r \times B^s)^{\Delta 0} = \sqrt{\frac{1}{2}} \left\{ (A^r(p) \times B^s(p))^{\Delta} + (A^r(n) \times B^s(n))^{\Delta} \right\}, \quad (32)$$

where Δ is the spin rank of the operator (and 0 refers to isoscalar). For $\Delta = 0$ we have, except for a factor of $\sqrt{2}$, the number operator as given in eq. (31).

The isovector, one-body operator has three possible projections. The $Z = -1$ component,

$$(A^r \times B^s)^{\Delta 1}_{-1} = (A^r(p) \times B^s(n))^{\Delta}, \quad (33a)$$

is related to β^+-decay or to (n,p) type charge exchange processes. The $Z = 0$ component,

$$(A^r \times B^s)^{\Delta 1}_{0} = \sqrt{\frac{1}{2}} \left\{ (A^r(p) \times B^s(p))^{\Delta} - (A^r(n) \times B^s(n))^{\Delta} \right\}, \quad (33b)$$

is used to express isovector electromagnetic transition operators. The $Z = +1$ component,

$$\left(A^r \times B^s \right)^{\Delta 1}_{+1} = \left(A^r(n) \times B^s(p) \right)^{\Delta}, \quad (33c)$$

is needed in β^--decay and (p,n) type of charge exchange processes. For more complicated operators, the isospin projection can be worked out using Clebsch-Gordan coefficients and the equivalence relationships given by eqs. (28).

VII.8   Summary of isospin projection

The trace of an operator in a subspace with definite isospin can be obtained in the following way starting from a set of traces calculated by treating protons and neutrons as distinguishable particles:

    (1) Reexpress the single-particle operators in terms of isospin operators using eqs. (28).

(2) Couple the single-particle operators to definite isospin ranks as done *e.g.*, in eqs. (32) and (33), and calculate the fixed-Z traces of these operators from proton-neutron traces.

(3) Use eq. (23) to project out the fixed-isospin traces from the fixed-Z traces.

Although we have derived the method in terms of scalar-T averaging, it is easy to see that it can be used also to obtain the configuration-T averages. Steps (1) and (2) are the same in both cases since only the operator is being manipulated. For step (3), we start with proton-neutron configurations as in section 2 for finding the dimension in mT-subspace, and calculate all the m_{pn} traces belonging to configuration m for a given Z-value. This gives the mZ-traces. The mT-traces can be projected from a set of these mZ traces in the manner expressed by eq. (27).

Chapter VIII

Spin and Isospin

Since spin and isospin are usually taken as good quantum numbers, the distribution of a quantity in the subspace of fixed J and T is of interest. However, calculations involving J and T are often time-consuming because of the angular momentum recouplings involved. In this chapter we shall describe three different methods to obtain fixed spin-isospin traces. Two of these methods involve the projection of fixed-J traces from fixed-M traces, in the same way as carried out for isospin in the previous chapter; the difference between these first two methods lies in the way the fixed-M traces are calculated. The third method uses techniques of spherical tensor algebra to reduce a fixed-JT trace in several orbits into a product of single-orbit traces. Not many studies have actually been carried out with fixed-JT averaging, perhaps because none of the available methods is very satisfactory. The best hope may well be in developing efficient approximations for the required traces.

VIII.1 Dimension of configuration-spin space

The most straightforward method to obtain fixed-J traces is to project them from a set of fixed-M traces in the same way as we have done in the previous chapter for isospin. Let us use the calculation of the number of states in the subspace of a given J value as an illustration. Consider first m identical particles in Λ active single-particle orbits. The single-particle states are labelled by j_r and μ_r, the projection of j_r on the quantization axis. For m_r particles in orbit r, the maximum value of the projection quantum number M is obtained by putting the particles in the single-particle states with the largest possible μ_r-values,

$$M_r(\text{max}) = j_r + (j_r - 1) + (j_r - 2) + \cdots + (j_r - m_r + 1)$$

$$= \frac{m_r(2j_r + 1 - m_r)}{2}. \tag{1}$$

The maximum M value for a configuration m is therefore

$$M_m(\text{max}) = \sum_r \frac{m_r(2j_r + 1 - m_r)}{2}. \tag{2}$$

The maximum J value for this configuration $J_m(\text{max})$ must be equal to $M_m(\text{max})$ since, if there were a higher J-value in the configuration, there would be states with higher M-values and this is not true. The subspace of maximum M is therefore the simplest subspace in a configuration having only one possible J value associated with it. The dimension of the $J_m(\text{max})$ subspace is therefore $(2J_m(\text{max}) + 1)d_{m,M(\text{max})}$ since $J_m(\text{max})$ can take on $(2J_m(\text{max}) + 1)$ different M values.

The number of states with lower J-values can be found from $d_{m,M}$, the number of states in configuration m with projection M, in an analogous manner as in eq. (VII-4),

$$d_{mJ} = (2J + 1)(d_{m,M=J} - d_{m,M=J+1}). \tag{3}$$

Given a set of single-particle states, each with a definite μ-value, the values of $d_{m,M}$ can be obtained easily by counting the number of states of a given M value that can be constructed in the configuration. Although a simple mathematical expression cannot be given for $d_{m,M}$, the counting itself is quite straightforward and can be carried out easily on a computer.

Let us now consider a mixture of protons and neutrons: each distribution of particles among the single-particle states is now characterized by M as well as Z, half the difference between the number of protons and the number of neutrons. Let the number of states for a given (M, Z)-value in configuration m be $d_{m,M,Z}$. The number of states for a given (J, Z) in this subspace can be obtained in the same way as in eq. (3),

$$d_{mJ,Z} = d_{m,M=J,Z} - d_{m,M=J+1,Z}.$$

From this expression we can obtain the number of states of a given (J, T) in configuration m,

$$\begin{aligned}
d_{mJ,T} &= (2T + 1)(d_{mJ,Z=T} - d_{mJ,Z=T+1}) \\
&= (2J + 1)(2T + 1)\Big\{ d_{m,M=J,Z=T} - d_{m,M=J+1,Z=T} \\
&\qquad\qquad - (d_{m,M=J,Z=T+1} - d_{m,M=J+1,Z=T+1})\Big\} \\
&= (2J + 1)(2T + 1)\Big\{ d_{m,M=J,Z=T} + d_{m,M=J+1,Z=T+1} \\
&\qquad\qquad - d_{m,M=J,Z=T+1} - d_{m,M=J+1,Z=T}\Big\}. \tag{4}
\end{aligned}$$

This is essentially the same as eq. (3) except that isospin projection is included.

VIII.2   Projection of spin-isospin traces

The calculation of the number of states in a given subspace can be viewed as taking the trace of the unity operator. The idea underlying eqs. (3) and (4) can also be applied to operators in general. Let the spin-isospin rank of the operator be represented by the single Greek letter Δ. If the trace of $\hat{O}^\Delta$ in the subspace of a given (M, Z)-value in configuration m is known, the configuration-JT trace can be obtained by a similar procedure as used in eq. (VII-27) for isospin alone. For a scalar operator, $\Delta = 0$, all the traces are independent of (M, Z), and the relation simplifies to

$$\ll \hat{O}^0 \gg^{(JT)} = (2J + 1)(2T + 1)$$
$$\times \{ \ll \hat{O}^0 \gg^{(M=J,Z=T)} + \ll \hat{O}^0 \gg^{(M=J+1,Z=T+1)}$$
$$- \ll \hat{O}^0 \gg^{(M=J,Z=T+1)} - \ll \hat{O}^0 \gg^{(M=J+1,Z=T)} \}, \quad (5)$$

in analogy with eq. (VII-23).

For non-scalar operators the projection must be done step by step starting from the maximum (J, T) value and using the Wigner-Eckart theorem to account for the (M, Z) dependences of the traces, in the same way as outlined in section VII.6 for isospin projection. The usefulness of the expression, however, depends on whether the (M, Z)-traces can be obtained easily.

VIII.3   Fixed-M traces

The method used by Jacquemin (1980) to find the fixed-M trace of an operator is based on a space made of single-particle states labelled by μ_i, the projection of j on the quantization axis. Let us start with a space spanned by N single-particle states each with a distinct μ_i-value. Such a space can be realized, e.g., in the spherical shell model basis by having only a single active j-orbit. In this simple case, $N = 2j + 1$ and $\mu_i = -j$, $-j + 1, -j + 2, \ldots, j$. However, a spherical basis is not necessary; in fact, one of the strengths of the method is that it can be applied equally well to a deformed shell model basis.

In the same way as in Chapter V, the operator is first decomposed according to particle rank k,

$$\hat{O}^\Delta = \sum_k \hat{O}^\Delta(k). \quad (6)$$

Since only the diagonal part of the operator can contribute to the trace, we can write the diagonal k-particle part of $\hat{O}^\Delta$ in the form

$$\hat{O}^\Delta_{\text{diag.}} = \sum_{\text{all } \mu} \mathcal{V}^\Delta(\mu_1, \mu_2, \ldots, \mu_k) \, a^\dagger_{\mu_1} a^\dagger_{\mu_2} \cdots a^\dagger_{\mu_k} a_{\mu_k} \cdots a_{\mu_2} a_{\mu_1}, \quad (7)$$

where the single-particle creation operator $a^\dagger_{\mu_i}$ and annihilation operator a_{μ_i} respectively creates and annihilates a single-particle with projection quantum number μ_i. The defining matrix elements of $\hat{O}^\Delta(k)$ is given by the diagonal matrix elements in the k-particle space,

$$\mathcal{V}^\Delta(\mu_1, \mu_2, \ldots, \mu_k) = \langle \mu_1, \mu_2, \ldots, \mu_k | \hat{O}^\Delta(k) | \mu_1, \mu_2, \ldots, \mu_k \rangle, \qquad (8)$$

where the k-particle wave function, $|\mu_1, \mu_2, \ldots, \mu_k\rangle$, is normalized and antisymmetrized and can be taken as a Slater determinant with the single-particle states $\mu_1, \mu_2, \ldots, \mu_k$ occupied.

The trace of an operator $\hat{O}^\Delta$ in the subspace of m particles with total projection M is then given by

$$\ll \hat{O}^\Delta \gg^{(mM)} = \sum_k \sum_{\text{all } \mu} d_{m,M}(\mu_1, \mu_2, \ldots, \mu_k) \mathcal{V}^\Delta(\mu_1, \mu_2, \ldots, \mu_k), \qquad (9)$$

where $d_{m,M}(\mu_1, \mu_2, \ldots, \mu_k)$ is the number of states in the (m, M)-subspace with each of the k single-particle states $\mu_1, \mu_2, \ldots, \mu_k$ occupied. One of the most important steps in this approach is therefore to find the value of $d_{m,M}(\mu_1, \mu_2, \ldots, \mu_k)$ efficiently.

It was demonstrated by Jacquemin and Spitz (1979) that an iterative method is the most convenient way to calculate $d_{m,M}(\mu_1, \mu_2, \ldots, \mu_k)$. Consider first a one-body operator,

$$\hat{O}^\Delta_{\text{diag.}}(k = 1) = \mathcal{V}(\mu) a^\dagger_\mu a_\mu,$$

and let $\Phi_{m,M}(\alpha)$ be one of the m-particle state with projection M and the label α distinguish between different m-particle states of the same M value. When $a^\dagger_\mu$ acts on $\Phi_{m,M}(\alpha)$, a state of $m + 1$ particles and projection $M + \mu$ is produced if the single-particle state μ in $\Phi_{m,M}(\alpha)$ is unoccupied, and 0 if the single-particle state μ is already occupied. Let η be the number of m-particle states with single-particle state μ unoccupied. By definition, we have

$$d_{m,M} = d_{m,M}(\mu) + \eta,$$

where $d_{m,M}(\mu)$ is the number of m-particle states with projection M and single-particle state μ occupied.

We shall now show that η is equal to $d_{m+1,M+\mu}(\mu)$, the number of $(m + 1)$-particle states having projection $M + \mu$ and single-particle state μ occupied. When a_μ acts on one of these $d_{m+1,M+\mu}(\mu)$ states, an m-particle state with projection M and single-particle state μ empty is produced. Hence

$$d_{m+1,M+\mu}(\mu) \leq \eta. \qquad (10)$$

Similarly, when $a^\dagger_\mu$ acts on one of the η states with particle number m and μ unoccupied, it will produce a state with $m + 1$ particles and projection $M + \mu$. Hence,

$$\eta \leq d_{m+1,M+\mu}(\mu). \qquad (11)$$

From eqs. (10) and (11), it follows that

$$\eta = d_{m+1,M+\mu}(\mu),$$

and

$$d_{m,M} = d_{m,M}(\mu) + d_{m+1,M+\mu}(\mu).$$ (12)

From this we obtain the recurrence relation for $k = 1$,

$$d_{m,M}(\mu) = d_{m-1,M-\mu} - d_{m-1,M-\mu}(\mu).$$ (13)

By starting with $m = 1$, and using the fact that $d_{0,0}(\mu) = 0$ and $d_{0,0} = 1$, we can find all the values of $d_{m,M}(\mu)$ required for propagating $k = 1$ operators by eq. (9).

Since each single-particle state here has a unique μ value, extension of the considerations to k-body operators is immediate. For an operator of arbitrary particle-rank k, eq. (13) can be generalized to

$$d_{m,M}(\mu_1,\mu_2,\ldots,\mu_{k-1},\mu_k) = d_{m-1,M-\mu_k}(\mu_1,\mu_2,\ldots,\mu_{k-1})$$
$$- d_{m-1,M-\mu_k}(\mu_1,\mu_2,\ldots,\mu_{k-1},\mu_k).$$ (14)

To make use of the recurrence relation eq. (14), it is necessary to start with $k = 1$ and obtain all the values of $d_{m,M}(\mu)$ for all M using eq. (13). Once we have these values for $k = 1$, we can use eq. (14) to produce the values of $d_{m,M}(\mu_1,\mu_2)$ for $k = 2$ and so on until we reach the highest k value required.

Up to now we have been assuming that each single-particle state in the space has a unique μ-value. In general, there may be several single-particle states in the space with the same μ value but distinguished by other labels such as the principal quantum number, orbital angular momentum, and spin. The same propagation relation among traces given by eqs. (9) and (14) can be maintained if we interpret the single-particle operators $a_\mu^\dagger$ and a_μ as creating and annihilating respectively a particle with projection μ without regard to the other quantum numbers. This can be done in a more logical fashion by defining a new type of *orbit* labelled only by μ. The single-particle states within such a μ-*orbit* are distinguished by other labels which need not be specified here. This is similar to the case of spherical orbits where each orbit is labelled by the principal quantum number, orbital angular momentum, and spin but leaving the projection μ unspecified. The difference is however an important one in that the various states in a spherical orbit are related to each other through a rotation around the quantization axis. No such simple relation exists for the μ-orbits defined here; however it does have the advantage of being applicable in a deformed basis.

Let the number of states in μ-orbit r be N_r, a number depending on the active space chosen, and let the total number of μ-orbits be Λ. If there is an underlying spherical basis, then $\Lambda = 2j_{\max} + 1$ where $j_{\max}$ is the largest single-particle spin in the space. A configuration in such a space is given by

$$\nu \equiv \{\nu_1, \nu_2, \ldots, \nu_\Lambda\},$$

where $\nu_r (\leq N_r)$ is the number of particles in the μ-orbit r.

A k-body operator can be written in a form analogous to that in eqs. (7) and (8),

$$\hat{O}^\Lambda_{\text{diag.}}(k) = \sum_\nu \mathcal{V}^\Lambda(k, \nu) \sum_{\mu' s \in \nu} a^\dagger_{\mu_1} a^\dagger_{\mu_2} \cdots a^\dagger_{\mu_k} a_{\mu_k} \cdots a_{\mu_2} a_{\mu_1}, \tag{15}$$

where the summation over μ_1 to μ_k is restricted to those values allowed by ν, and

$$\mathcal{V}^\Lambda(k, \nu) = \ll \hat{O}^\Lambda(k) \gg^{(k, \nu)}. \tag{16}$$

The propagation of the trace from the k-particle space to the m-particle space can be carried out in a way similar to eq. (9),

$$\ll \hat{O}^\Lambda \gg^{(m, M)} = \sum_{k, \nu} d_{m,M}(k, \nu) \, \mathcal{V}^\Lambda(k, \nu). \tag{17}$$

Here, $d_{m,M}(k, \nu)$ is to be interpreted as the number of times the k-particle configuration ν occurs in the subspace mM and is given by the same recurrence relation as in eq. (14).

For a scalar elementary operator, and product operators made of the powers of scalar elementary operators, Verbaarschot and Brussaard (1981) have shown that it is possible to use a generalized Wick theorem in a way similar to the diagrammatic method described in Chapter V to evaluate the fixed-M trace. For example, using the μ-orbit basis as described above, we have

$$\ll H^p \gg^{(mM)} = \sum_{\text{all } p} \sum_{t_i=0}^{p_i} \sideset{}{'}\sum_{m_i} \binom{N - p_i}{m_i - t_i} D^{p_i}_{t_i}(H^p), \tag{18}$$

where the summation over m_i, the number of particles in orbit μ_i, is restricted by the conditions

$$\sum_i m_i = m, \quad \text{and} \quad \sum_i m_i \mu_i = M.$$

The symbol $D^{p_i}_{t_i}(H^p)$ represents all possible contractions of H^p with a total of p_i contractions in the μ-orbit i, and t_i of these contractions being left-contractions. The method is convenient for taking traces of product operators made up of powers of the same scalar elementary operator, such as H. As shown in Chapter V, if the elementary operators are *inequivalent*, there is the "counting" problem that was already complicated without μ; it may therefore prove to be difficult to extend the method to a more general class of product operators.

VIII.4   Approximate methods to evaluate fixed-M traces

In a space made up of a single spherical j-orbit, the allowed μ-values are $-j, -j+1, \ldots, j$, evenly spaced and distributed symmetrically around $\mu = 0$. With m particles, the possible $M(= \sum_i \mu_i)$ value ranges from $-M(\text{max})$ to $+M(\text{max})$, with $M(\text{max})$ given by eq. (2). The number of many-particle states for a given M value is also symmetrically distributed around $M = 0$ but with a maximum at $M = 0$ since there are many more different arrangements of the m particles to arrive at low $|M|$ values than at high $|M|$ values. In the absence of the Pauli principle, which prevents two particles to take on the same μ value in the single j-orbit case here, the distribution can be shown to be Gaussian for large j values. For dilute systems, $(2j + 1) \gg m \gg 1$, the Pauli principle is not important since the probability for two particles to occupy the same single-particle state is small. As a result, we can expect the distribution of the number of states as a function of M to be asymptotically Gaussian. The same must also be true in the space of several orbits as long as the system is dilute.

It is not crucial for the M-distribution to be exactly Gaussian to use the approximation method (Haq and Wong, 1979) described below. As mentioned earlier, it is fairly simple to calculate $d_{m,M}$, the dimension of a subspace with a given M value in the m-particle space. With a complete set of $d_{m,M}$ for all possible values of M for a given m, the moments of the M-distribution can be obtained. For order k, we have

$$\langle J_z^k \rangle^m = \frac{1}{d_m} \sum_M d_{m,M} M^k, \tag{19}$$

where $d_m = \sum_M d_{m,M}$. Since the distribution is symmetric around $M = 0$, all the odd moments vanish.

With the moments available to arbitrarily high order, we can obtain polynomials $P_\nu(M)$ of M by eq. (III-29). With these polynomials, an expansion of the traces of an operator in the mM-subspace can be made in terms of traces in the m-particle space using eq. (III-31)

$$\langle \hat{O}^\Delta \rangle^{(mM)} = \sum_\nu \langle \hat{O}^\Delta P_\nu(J_z) \rangle^{(m)} P_\nu(M). \tag{20}$$

The only difference here is that, instead of energy, the expansion is made in terms of M. Since the M-distribution is almost Gaussian, $P_\nu(x) \approx (\nu!)^{-\frac{1}{2}} He_\nu(x)$, where $He_\nu(x)$ is the Hermite polynomial of order ν, and we expect the series in eq. (20) to be rapidly convergent. This is a necessary criterion for the success of the approximation scheme since the higher-order polynomial traces $\ll \hat{O}^\Delta P_\nu(J_z) \gg^{(m)}$ are difficult to calculate.

For dilute systems, eq. (20) is a very attractive method to obtain fixed-M traces, and thence fixed-J traces via eq. (5). The main calculations involved are the scalar traces of $(\hat{O}^\Delta J_z^k)$ which can be easily carried out as long as k is not too large. A further advantage of the method is that we can replace the scalar averages by configuration averages in eqs. (19) and (20) and thus obtain the mJ-averages of $\hat{O}^\Delta$.

Good accuracy is essential for the M-expansions in eq. (20) since we have to take the differences between two such series to obtain the fixed-J moments that are of ultimate interest. The cumulative errors can be serious since we wish to truncate the series to as low orders as possible.

A variant of the method was proposed by Halemane (1981). Instead of M, the variable $J(J+1)$ and, instead of J_z, the operator J^2 is used in the expansion in eq. (20). The difficulty here is that $J(J+1)$ is a non-negative number and the number of states for a given J-value peaks at some low J-values, not necessarily the lowest one in the m-particle space. An underlying Gaussian distribution is therefore absent except in the M-distribution described above. The convergence of an expansion in $J(J+1)$ therefore has no guarantee of its own except by going back to the M-distribution.

VIII.5   Decomposition of fixed-JT traces into single-orbit traces

A method to calculate fixed-JT traces directly without going through the (M, Z)-traces is also in use (Mugambi, 1970, Lougheed and Wong, 1975). In a space consisting of Λ spherical j-orbits, the m-particle wave function for a given JT-state can be written as the angular momentum coupled product of single-orbit wave functions. In order to simplify the notation, we shall again use a single Greek letter to stand for both spin and isospin. Thus, for angular momentum recoupling coefficients and statistical factors, each symbol stands for a product of two such factors, one for spin and one for isospin.

In each orbit, the m_r particles are coupled to angular momenta γ_r. All other labels are represented by x_r, which need not be specified since our interest is in the trace over m-particle states with all possible x_r values. The single-orbit wave functions $\phi^{\gamma_r}(x_r)$ for different orbits are coupled together to give the final spin-isospin Γ of the $m = (\sum_r m_r)$-particle wave function. There are many possible ways to arrange the angular momentum coupling and for later convenience we shall take the one, sometimes referred to as the fan-shaped coupling scheme, arranged in the following manner,

$$\Phi^{JT}(m\gamma\Gamma;x) = \left(\cdots \left(\left(\phi^{\gamma_1}(x_1) \times \phi^{\gamma_2}(x_2) \right)^{\Gamma_2} \times \phi^{\gamma_3}(x_3) \right)^{\Gamma_3} \right.$$

$$\left. \times \cdots \times \phi^{\gamma_A}(x_A) \right)^{\Gamma}, \quad (21)$$

with the understanding that

$$\Gamma_A \equiv \Gamma \equiv (J, T), \qquad \text{and} \qquad \Gamma_1 \equiv \gamma_1.$$

In other words, we always couple the first orbit to the second one with intermediate coupling angular momentum given by the vector addition $\Gamma_2 = \gamma_1 + \gamma_2$. The resultant is then coupled to the third orbit with intermediate angular momentum given by the vector addition $\Gamma_3 = \Gamma_2 + \gamma_3$, and so on till we come to the last orbit.

By the same token, the operator $\hat{O}^\Delta$ can also be decomposed into a product of single-orbit operators $\hat{O}^{\omega_r}$ coupled together in the same manner as in eq. (21) for the wave functions,

$$\hat{O}^\Delta = \sum_\omega \mathcal{V}(\omega\Delta)\left(\cdots \left((\hat{O}_1^{\omega_1} \times \hat{O}_2^{\omega_2})^{\Delta_2} \times \hat{O}_3^{\omega_3} \right)^{\Delta_3} \times \cdots \times \hat{O}_A^{\omega_A} \right)^\Delta. \quad (22)$$

Each $\hat{O}_r^{\omega_r}$ is made up of the angular-momentum coupled product of single-particle creation and annihilation operators of a given orbit only and arranged in normal order. All other properties of the operator except for the angular momentum structure are contained in the defining matrix element $\mathcal{V}(\omega\Delta)$. The reduced matrix element of the product of single-orbit operators in eq. (22) can be transformed into a product of single-orbit reduced matrix elements using standard spherical tensor reduction techniques given in Appendix C. For traces, the situation is further simplified since we need only the diagonal terms.

Let us illustrate the procedure with the first $(A-1)$ orbits grouped together as a single unit represented by $\Phi^{\Gamma_{A-1}}$ for the state function and $\hat{O}^{\Delta_{A-1}}$ for the operator. We can isolate the contributions from orbit A, the last one in the product, from the multi-orbit reduced matrix element using eq. (C-23),

$$\langle (\Phi^{\Gamma_{A-1}} \times \phi^{\gamma_A})^\Gamma \| (\hat{O}^{\Delta_{A-1}} \times \hat{O}^{\omega_A})^\Delta \| (\Phi^{\Gamma_{A-1}} \times \phi^{\gamma_A})^\Gamma \rangle$$

$$= [\Gamma][\Delta]^{\frac{1}{2}} \left\{ \begin{matrix} \Gamma_{A-1} & \gamma_A & \Gamma \\ \Gamma_{A-1} & \gamma_A & \Gamma \\ \Delta_{A-1} & \omega_A & \Delta \end{matrix} \right\} \langle \Phi^{\Gamma_{A-1}} \| \hat{O}^{\Delta_{A-1}} \| \Phi^{\Gamma_{A-1}} \rangle \langle \phi^{\gamma_A} \| \hat{O}^{\omega_A} \| \phi^{\gamma_A} \rangle.$$

By applying eq. (C-23) $(\Lambda - 2)$ more times, each time removing the contributions of the last orbit from the multi-orbit reduced matrix element remaining from the previous application, we arrive at the result

$$\langle \Phi^{JT}(m\gamma\Gamma;x) \| \hat{O}^{\Delta} \| \Phi^{JT}(m\gamma\Gamma;x) \rangle$$

$$= \sum_{\omega} \mathcal{V}(\omega\Delta) \left\langle \left(\cdots \left((\phi^{\gamma_1}(x_1) \times \phi^{\gamma_2}(x_2))^{\Gamma_2} \times \phi^{\gamma_3}(x_3) \right)^{\Gamma_3} \right. \right.$$

$$\left. \times \cdots \times \phi^{\gamma_\Lambda}(x_\Lambda) \right)^{\Gamma} \left\| \left(\cdots \left((\hat{O}_1^{\omega_1} \times \hat{O}_2^{\omega_2})^{\Delta_2} \times \hat{O}_3^{\omega_3} \right)^{\Delta_3} \times \cdots \hat{O}_\Lambda^{\omega_\Lambda} \right)^{\Delta} \right.$$

$$\left\| \left(\cdots \left((\phi^{\gamma_1}(x_1) \times \phi^{\gamma_2}(x_2))^{\Gamma_2} \times \phi^{\gamma_3}(x_3) \right)^{\Gamma_3} \times \cdots \times \phi^{\gamma_\Lambda}(x_\Lambda) \right)^{\Gamma} \right\rangle$$

$$= \sum_{\omega} \mathcal{V}(\omega\Delta) \langle \phi^{\gamma_1} \| \hat{O}^{\omega_1} \| \phi_1^{\gamma} \rangle \prod_{i=2}^{\Lambda} [\Gamma_i][\Delta_i]^{\frac{1}{2}} \left\{ \begin{matrix} \Gamma_{i-1} & \gamma_i & \Gamma_i \\ \Gamma_{i-1} & \gamma_i & \Gamma_i \\ \Delta_{i-1} & \omega_i & \Delta_i \end{matrix} \right\} \langle \phi^{\gamma_i} \| \hat{O}^{\omega_i} \| \phi_i^{\gamma} \rangle.$$

$$(23)$$

Note that the right-hand side of the equation has only single-orbit reduced matrix elements.

The angular momentum structure of a matrix element is not affected by a summation over all the non-angular momentum labels x of all the orbits. Hence the reduction into single-orbits given by eq. (23) carries over to traces as well,

$$\ll \hat{O}^{\Delta} \gg^{(m\gamma\Gamma)} = \sum_{\omega} \mathcal{V}(\omega\Delta) \ll \hat{O}^{\omega_1} \gg^{(m_1\gamma_1)}$$

$$\times \prod_{i=2}^{\Lambda} [\Gamma_i][\Delta_i]^{\frac{1}{2}} \left\{ \begin{matrix} \Gamma_{i-1} & \gamma_i & \Gamma_i \\ \Gamma_{i-1} & \gamma_i & \Gamma_i \\ \Delta_{i-1} & \omega_i & \Delta_i \end{matrix} \right\} \ll \hat{O}^{\omega_i} \gg^{(m_i\gamma_i)} . \quad (24)$$

We have therefore a way to reduce a multi-orbit trace of an arbitrary operator into a product of single-orbit traces of operators $\hat{O}^{\omega_i}$. Since the types of single-orbit operators are limited, it is conceivable to have all the required traces calculated once and for all, and stored for later use. Indeed, some of this work has already been done, and the results have been applied to test cases with encouraging results (Lougheed and Wong, 1975; Wong and Lougheed, 1978; Ng and Wong, 1976).

The success of the method depends on whether the single-orbit traces can be calculated easily and stored in a convenient way for later retrieval. One of the methods to calculate these single-orbit traces is given in Appendix F where the traces of single-orbit operators are propagated from the defining space of the operator. Since, by definition, the operators are products of single-particle creation and annihilation operators, their traces are purely *geometrical* factors related only to the structure of the space of a single j-orbit: the physics enters only at the stage of eq. (24) after bringing in the defining matrix elements $\mathcal{V}(\omega\Delta)$. However, since spin and isospin are involved, the propagation of the single-orbit traces requires angular momentum recoupling and, as a result, the calculations become very time consuming for large j values.

Some simplification of the method may be possible. In orbits with large j values, the space is almost classical in the sense that both the discrete nature of the distribution of states and the Pauli principle are taking on less important roles. It is possible then that approximations can be found by which exact calculations of the recoupling coefficients may be unnecessary. This may also resolve some of the difficulties in storing the large number of traces for the large j-value orbits. If the approximation is sufficiently fast for some of the traces, they can be calculated when required rather than retrieved from storage.

A more severe problem with the method concerns the amount of information generated. The space is now subdivided into a very fine mesh specified by a set of $(m\gamma\Gamma)$-values. The dimension of such a subspace is the product of the number of states in each orbit differing only in the non-angular momentum label x_r. For most cases, x_r is unity in a given orbit. For example, in jj-coupling, x_r represents the labels (s,t), the seniority and reduced isospin, and the multiplicity of the (s,t) representation for a given J, T (Hamermesh, 1962). Since we are calculating the traces in a given $(m\gamma\Gamma)$-subspace, we have a distribution of the physical quantity of interest in each of the subspaces. It is not possible, in general, to make use of the large amount of detailed information generated in the process, and one often ends up discarding most of it by summing over many such subspaces. For example, the intermediate coupling angular momenta Γ are for most purposes devoid of physical significance, and it incurs no loss of useful information by summing them away. It would be useful if a way could be found to avoid these intermediate stages of recoupling involving the values of Γ and thus reduce the number of time-consuming intermediate state angular momentum recoupling coefficients in eq. (24). However, no such prospect seems to be in the offing.

Chapter IX

Irreducible Representations of Groups

So far we have been concerned with averaging operators in the m-particle space and its subspaces defined according to configuration m, spin J, and isospin T. In every case there is a definite group structure associated with the states over which the traces are taken. For example, in scalar averaging all the states belong to a particular irreducible representation of $U(N)$, the unitary group of N dimensions, and in configuration averaging they belong to $\sum_r U(N_r)$, the direct sum subgroup of $U(N)$. For spin and isospin, the groups associated with the subdivision are $R(3)$ and $U(2)$ respectively.

The methods used to calculate the traces are based on propagation, *i.e.*, relating the traces in one space or subspace to those in another. For such propagation relations to exist, an underlying group structure among the states is essential. On the other hand, the existence of a group structure alone does not imply that the traces can be propagated. As we shall see later in this chapter, for a large number of groups of interest it is not possible to propagate the traces from one irreducible representation to another, except for simple operators. We could have considered methods to obtain traces other than those based on propagation; however, for the high dimensional spaces we are interested in here, they are usually too cumbersome to be practical.

In this chapter, we shall first briefly reformulate some of the averaging techniques given in earlier chapters in a more group theoretical language to set the stage for a discussion of averaging over other group representations. We shall not, however, attempt to give an account of group symmetry in nuclear structure for which excellent and concise accounts exist, *e.g.*, in Parikh (1978), nor shall we give the mathematics that is well described in standard references such as Hamermesh (1962).

IX.1 Averaging over $U(N)$

Let us use here the vector m to represent a particular distribution of m particles over the N single-particle states with the occupancy in each being 0 or 1 to satisfy the Pauli principle. This usage of m is similar to the configuration averaging case except that here the vector is in the space of N single-particle states rather than in the space of Λ orbits. A different arrangement of the m particles in the N single-particle states, $\boldsymbol{m'}$, is related to $\boldsymbol{m}$ through a transformation of the form

$$\boldsymbol{m'} = \mathcal{T}_{m,m'}\,\boldsymbol{m}. \tag{1}$$

An element of the transformation matrix $\mathcal{T}$ can be written as the product of m operators each of which has the form

$$u_{ij} = a_i^\dagger a_j\,, \tag{2}$$

where $a_i^\dagger$ creates a particle in the single-particle state i, and a_j annihilates one in the single-particle state j. Using the anticommutation relation

$$\{a_i^\dagger, a_j\} = \delta_{ij}\,,$$

we obtain the relation

$$[u_{ij}, u_{kl}] = u_{il}\delta_{jk} - u_{kj}\delta_{il}\,. \tag{3}$$

Consequently the N^2 quantities u_{ij} form a closed algebra and satisfy the requirements to be the infinitesimal operators of a Lie group. Furthermore, since the lengths of all the vectors $\boldsymbol{m}$ for a given particle number m are the same, $\mathcal{T}$ is unitary ($\mathcal{T}\mathcal{T}^\dagger = \mathcal{T}^\dagger\mathcal{T} = 1$), and all the states represented by $\boldsymbol{m}$ belong to the $[1^m]$ representation of $U(N)$.

The trace of a k-body operator $\hat{O}(k)$ in the m-particle space can be calculated using a slightly different approach from that used in Chapter IV. There is only one Casimir invariant operator, or scalar, that commutes with all the generators in the group, and it can be taken to be the number operator,

$$\hat{n} = \sum_{i=1}^{N} a_i^\dagger a_i\,.$$

Obviously $\langle m|\hat{n}|m'\rangle = m\,\delta_{mm'}$.

Since only the part of $\hat{O}(k)$ that is invariant under a unitary transformation of the space can contribute to the trace, the *trace-equivalent* form, $\hat{O}_{\mathrm{tr}}(k)$, of $\hat{O}(k)$ must be a polynomial in the invariant operator, $\hat{n}$ in this case. At the same time, for $\hat{O}_{\mathrm{tr}}(k)$ to be a k-body operator, it must vanish for all $0 \le m < k$. The trace-equivalent operator for $\hat{O}(k)$ thus takes the form

$$\hat{O}_{\mathrm{tr}}(k) \rightarrow \langle \hat{O}(k)\rangle^{(k)}\,(\hat{n}-k+1)(\hat{n}-k+2)\cdots(\hat{n}-1)\hat{n} \rightarrow \langle \hat{O}(k)\rangle^{(k)} \binom{\hat{n}}{k},$$

where we have taken out a factor $k!$ to arrive at the final result. The defining quantity for $\hat{O}(k)_{\text{tr}}$ is given by $\langle\hat{O}(k)\rangle^{(k)}$, the trace of $\hat{O}(k)$ in the k-particle space. In terms of average traces, we obtain

$$\langle\hat{O}(k)\rangle^{(m)} = \langle\hat{O}(k)\rangle^{(m)} = \binom{m}{k}\langle\hat{O}(k)\rangle^{(k)},$$

the same result as eq. (IV-14).

The simplicity in averaging over the $U(N)$ group derives from the fact that each irreducible representation is uniquely characterized by m. Hence any operator of a definite particle rank is given by a single defining quantity and the average of this operator is propagated by a function of $\hat{n}$ alone. As we shall see later, it is essential to have enough invariant operators in order to propagate the averages among the irreducible representations in a particular group.

IX.2   Direct sum subgroups

When the N single-particle states are partitioned into orbits, with N_r single-particle states in orbit r, the group $U(N)$ is said to be decomposed into a direct sum subgroup $U(N_r)$

$$U(N) = \sum_r U(N_r). \tag{4}$$

It is common practice to use a spherical basis for the subdivision but this is not necessary for the discussion here. The only requirement is that there is no mixing of single-particle states belonging to different orbits under unitary transformations.

An irreducible representation of the direct sum subgroup is characterized by the number of particles in each orbit. Similarly, we can define a particle rank k_r for each orbit and an operator of definite particle ranks in all the orbits can be written as $\hat{O}(k)$. In each orbit, we have an invariant operator $\hat{n}_r$ that is independent of a rotation among the states in the orbit. The trace-equivalent form of $\hat{O}(k)$ is then

$$\hat{O}_{\text{tr}}(k) = \left\{\prod_{r=1}^{\Lambda}\binom{\hat{n}_r}{k_r}\right\}\langle\hat{O}(k)\rangle^{(k)} = \binom{\hat{n}}{k}\langle\hat{O}(k)\rangle^{(k)},$$

and the configuration average is propagated by the relation

$$\langle\hat{O}(k)\rangle^{(m)} = \binom{m}{k}\langle\hat{O}(k)\rangle^{(k)},$$

where $\langle\hat{O}(k)\rangle^{(k)}$ is the defining average for $\hat{O}_{\text{tr}}(k)$. In terms of traces, we have

$$\ll\hat{O}(k)\gg^{(m)} = \binom{N-k}{m-k}\ll\hat{O}(k)\gg^{(k)},$$

the analogy to eq. (V-19) for a k-body operator.

IX.3   Direct product subgroups

If a space contains both protons and neutrons, the states in it can be classified according to isospin T. To make use of the isospin symmetry, it is convenient to label each single-particle creation and annihilation operator by two indices: one index, $(r, s, \dots)$, for the single-particle orbit it operates in, and another one, (α, β), to indicate whether it is a proton or a neutron. Instead of eq. (2), the N^2 generators of the $U(N)$ group can now be written in the form

$$u_{r\alpha,s\beta} = a^\dagger_{r\alpha} a_{s\beta} , \tag{5}$$

and with these we can construct the following two types of generators:

$$U_{rs} = \sum_\alpha u_{r\alpha,s\alpha} , \tag{6a}$$

$$U_{\alpha\beta} = \sum_r u_{r\alpha,r\beta} , \tag{6b}$$

satisfying the commutation relations

$$[U_{rs}, U_{tu}] = U_{ru}\delta_{st} - U_{ts}\delta_{ru}, \qquad [U_{\alpha\beta}, U_{\gamma\epsilon}] = U_{\alpha\epsilon}\delta_{\beta\gamma} - U_{\gamma\beta}\delta_{\alpha\epsilon},$$

while the commutators between generators of two groups vanish,

$$[U_{rs}, U_{\alpha\beta}] = 0,$$

as can be seen by substituting eqs. (6) into eq. (3).

The $(N/2)^2$ generators given by eq. (6a) form the group $U(N/2)$, and the four generators given by eq. (6b) form the group $U(2)$. Thus we have made a decomposition of $U(N)$ into a direct product of $U(N/2)$ and $U(2)$,

$$U(N) = U(N/2) \times U(2). \tag{7}$$

In addition to the number operator $\hat{n}$, we now have an additional invariant T^2 of $U(2)$, which can be used to propagate the traces of the irreducible representations of $U(N/2) \times U(2)$ as carried out in Chapter VII.

For configuration-isospin averaging, the decomposition is

$$U(N) = \sum_r U(N_r/2) \times U(2).$$

The invariants are then the particle number operator $\hat{n}_r$ of each orbit and T^2, the square of the isospin operator.

IX.4   Averaging over states of arbitrary symmetry

Before considering a trace calculation in general it is useful to review here
the usual approach for obtaining the matrix elements of an operator. For
simplicity, we shall only consider a number-conserving operator $\hat{O}(k)$ of
particle rank k which can be written in the form

$$\hat{O}(k) = \sum_{\substack{\alpha'_1, \alpha'_2, \cdots, \alpha'_k \\ \alpha_1, \alpha_2, \cdots, \alpha_k}} \langle \alpha'_1, \alpha'_2, \cdots, \alpha'_k | \hat{O} | \alpha_1, \alpha_2, \cdots, \alpha_k \rangle$$

$$\times a^{\dagger}_{\alpha'_1} a^{\dagger}_{\alpha'_2} \cdots a^{\dagger}_{\alpha'_k} a_{\alpha_1} a_{\alpha_2} \cdots a_{\alpha_k}, \qquad (8)$$

where $\langle \alpha'_1, \alpha'_2, \cdots, \alpha'_k | \hat{O} | \alpha_1, \alpha_2, \cdots, \alpha_k \rangle$ are the defining matrix elements.
A matrix element of $\hat{O}(k)$ in the m-particle space is given by

$$\langle \beta'_1, \beta'_2, \cdots, \beta'_m | \hat{O} | \beta_1, \beta_2, \cdots, \beta_m \rangle$$

$$= \sum_{\substack{\alpha'_1, \alpha'_2, \cdots, \alpha'_k \\ \alpha_1, \alpha_2, \cdots, \alpha_k}} \langle \alpha'_1, \alpha'_2, \cdots, \alpha'_k | \hat{O} | \alpha_1, \alpha_2, \cdots, \alpha_k \rangle$$

$$\times \langle \beta'_1, \beta'_2, \cdots, \beta'_m | a^{\dagger}_{\alpha'_1} a^{\dagger}_{\alpha'_2} \cdots a^{\dagger}_{\alpha'_k} a_{\alpha_1} a_{\alpha_2} \cdots a_{\alpha_k} | \beta_1, \beta_2, \cdots, \beta_m \rangle. \quad (9)$$

This expression can be interpreted as the *propagation* of the matrix elements of $\hat{O}(k)$ from the k-particle (defining) space to the m-particle space.
The propagators here are the matrix elements of products of single-particle
creation and annihilation operators. They are purely geometric factors and,
in the simplest case, are given by the occupancies of the single-particle
states. In this sense, they are similar in nature to the propagators for
traces.

The calculation of a trace is simpler than that given in eq. (9). In the
first place, only the diagonal part of $\hat{O}(k)$ is required. In the second place,
and of more importance, we are no longer interested in the individual states;
instead we are considering all the states in a subspace as a single unit. As
a result, instead of single-particle creation and annihilation operators, we
can use the group generators to express the operator involved in a trace
calculation. For a given group with irreducible representations labelled by
$[f]$, we can define a *trace-equivalent* operator $\hat{O}_{\mathrm{tr}}(k)$ for $\hat{O}(k)$ by

$$\ll \hat{O}_{\mathrm{tr}}(k) \gg^{(m[f])} = \ll \hat{O}(k) \gg^{(m[f])}, \qquad (10)$$

i.e., $\hat{O}_{\mathrm{tr}}(k)$ and $\hat{O}(k)$ give identical traces for all the irreducible representations of the group. We can express $\hat{O}_{\mathrm{tr}}(k)$ in the form

$$\hat{O}_{\mathrm{tr}}(k) = \sum_i \langle \hat{O}(k) \rangle^{(k[i])} P_i(u), \qquad (11)$$

where $\langle \hat{O}(k) \rangle^{(k[i])}$ is the defining quantity of $\hat{O}_{\mathrm{tr}}(k)$ for the group representation, and $P_i(u)$ is a polynomial in the generators u.

Eq. (11) also allows us in principle to propagate operators with nonzero tensorial ranks; however, for simplicity, we shall discuss here only scalar operators. In this case, only the invariants of the group can appear in the polynomials $P_i(u)$ in eq. (11). From eq. (11) we see that the number of defining averages for $\hat{O}_{\mathrm{tr}}(k)$ is given by the number of irreducible representations for particle number k. Consequently, one of the necessary conditions to propagate the traces of a k-body operator among the irreducible representations of a group is to have the number of linearly independent group invariants equal to the number of irreducible representations for k particles.

For low particle-rank operators, the Casimir operators alone are often adequate to propagate the traces. For example, as we have seen earlier, for $U(N)$ the traces of an operator of particle rank k are propagated by a single defining quantity, and the propagator $P_i(u)$ is a polynomial of rank k constructed out of the Casimir operator $\hat{n}$.

For $U(N/2) \times U(2)$, we have available the additional Casimir operator T^2 from the isospin $U(2)$ group. In the one-particle space, there is only one irreducible representation, labelled by $(m, T) = (1, \frac{1}{2})$, and the two invariants are equivalent in this case since the one-body part of T^2 is proportional to $\hat{n}$. Consequently, the propagation of a one-body operator involves only a single term as we have seen earlier in section VII.3. For two particles, there are two irreducible representations, $(m, T) = (2, 0)$ and $(2, 1)$. The two defining averages of a two-body operator, $\hat{O}(2)$, can be taken to be $\langle \hat{O}(2) \rangle^{(2,0)}$ and $\langle \hat{O}(2) \rangle^{(2,1)}$. By inspection we can write

$$\hat{O}_{\mathrm{tr}}(2) = \left\{ \tfrac{1}{8} \hat{n}(\hat{n} + 2) - \tfrac{1}{2} T^2 \right\} \langle \hat{O}(2) \rangle^{(2,0)}$$

$$+ \left\{ \tfrac{3}{8} \hat{n}(\hat{n} - 2) + \tfrac{1}{2} T^2 \right\} \langle \hat{O}(2) \rangle^{(2,1)}, \qquad (12)$$

as the trace-equivalent operator for $\hat{O}(2)$. This expression is similar to that given by eq. (VII-11) for the Hamiltonian operator.

For $k = 4$, the maximum particle rank encountered in H^2, there are three defining averages corresponding to the three irreducible representations $(m, T) = (4, 2)$, $(4, 1)$, and $(4, 0)$. From the two Casimir operators $\hat{n}$ and T^2, only two independent $k = 4$ operators, $\hat{n}^4$ and $\hat{n}^2 T^2$, can be formed. As shown in eq. (VII-17), we can use T^4 as the additional invariant, and an expression for $\hat{O}_{\mathrm{tr}}(4)$ involving $\langle \hat{O}(4) \rangle^{(4,2)}$, $\langle \hat{O}(4) \rangle^{(4,1)}$, and $\langle \hat{O}(4) \rangle^{(4,0)}$ can be constructed in a similar manner as eq. (12). However, this expression is of little practical use since, as discussed in section VII.4, it is more convenient to use particle-hole transformations and replace the four-particle averages in the input with simpler ones involving only $m = 2$ averages as done in eq. (VII-21).

IX.5 $U(4)$ and the supermultiplet scheme

The decomposition of $U(N)$ into $U(N/4) \times U(4)$ is interesting because of possible $U(4)$ symmetry in nuclei. Wigner (1937) suggested that if the nuclear force is independent of both spin S and isospin T, a supermultiplet of states having the same $U(N/4)$ symmetry will be degenerate in energy. Since there is only a single $U(4)$ representation that can be coupled with that of $U(N/4)$ to form the completely antisymmetrized m-particle representation of $U(N)$, representations of $U(N/4)$ and $U(4)$ are conjugates of each other. The symmetry is also known as the space symmetry since $U(N/4)$ labels the space part of the wave functions. That is, if $U(N/4)$ is represented by a Young tableau $[f]$ of $N/4$ rows in the form

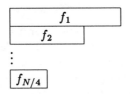

with the length of each row f_i restricted by the conditions

$$f_i \geq f_j \geq 0 \quad \text{for} \quad j > i, \quad \text{and} \quad \sum_i f_i = m,$$

then $U(4)$ is represented by $[\tilde{f}]$, obtained from $[f]$ by interchanging the rows and columns. In practice, the $U(N/4)$ symmetry is not exact since the nuclear force is spin-dependent and, to a much lesser extent, charge-dependent. However, in light nuclei there is strong evidence of the presence of $U(N/4)$ symmetry in both energy spectra and transition rates.

The four ST-states of $U(4)$ for a particle may be labelled by the projections of S and T on the quantization axes,

$$(m_s, m_t) = (+\tfrac{1}{2}, +\tfrac{1}{2}), \quad (+\tfrac{1}{2}, -\tfrac{1}{2}), \quad (-\tfrac{1}{2}, +\tfrac{1}{2}), \quad (-\tfrac{1}{2}, -\tfrac{1}{2}) \cdot$$

Each single-particle creation operator $a_i^\dagger$ and annihilation operator a_i is now distinguished by two indices: one, $(r, s, \ldots)$, for the space symmetry and another one, $(\phi, \chi, \psi, \omega)$, for the four (m_s, m_t) states. Analogous to eq. (6b), we can write the 16 generators of $U(4)$ in the form

$$U_{\phi\chi} = \sum_r a_{r\phi}^\dagger a_{r\chi} \,. \tag{13}$$

By inspection, we have the commutation relation

$$[U_{\phi\chi}, U_{\psi\omega}] = U_{\phi\omega}\delta_{\chi\psi} - U_{\chi\psi}\delta_{\phi\omega} \,, \tag{14}$$

among the generators.

Instead of $U(4)$, we can deal with $SU(4)$ constructed from 15 of the 16 generators of $U(4)$ by omitting the number operator. This is equivalent to ignoring transformations that are simply a change of the overall phase factor or scale. There are three Casimir invariants in $SU(4)$, a quadratic one, a cubic one, and a quartic one. They can be written respectively as

$$G_2(SU(4)) = \sum_{\chi\phi} U_{\phi\chi}U_{\chi\phi},$$

$$G_3(SU(4)) = \sum_{\chi\phi\psi} U_{\phi\chi}U_{\chi\psi}U_{\psi\chi},$$

$$G_4(SU(4)) = \sum_{\phi\chi\psi\omega} U_{\phi\chi}U_{\chi\psi}U_{\psi\omega}U_{\omega\phi}.$$

These three, together with the number operator $\hat{n}$, the Casimir operator for $U(N/4)$, are adequate to propagate low particle-rank operators in the supermultiplet scheme. However, as we shall see later, other group invariants must be found for the higher particle-rank operators.

The expectation value of the Casimir operators in a given $SU(4)$ irreducible representation $[f]$ can be expressed in terms of f_i (Parikh, 1973), the lengths of the four rows of the Young tableau representing the symmetry,

$$\langle [f] | G_2(SU(4)) | [f] \rangle = \sum_{i=1}^{4} f_i^2 + 3f_1 + f_2 - f_3 - 3f_4 , \tag{15a}$$

$$\langle [f] | G_3(SU(4)) | [f] \rangle = \sum_{i=1}^{4} f_i^3 + 6f_1^2 + 3f_2^2 - 3f_4^2$$
$$- \sum_{j>i} f_i f_j + 9f_1 - f_2 - 5f_3 - 3f_4 , \tag{15b}$$

$$\langle [f] | G_4(SU(4)) | [f] \rangle = \sum_{i=1}^{4} f_i^4 + 9f_1^3 + 5f_2^3 + f_3^3 - 3f_4^3 - \sum_{j>i}(f_i^2 f_j + f_i f_j^2)$$
$$+ 27f_1^2 + 5f_2^2 - 5f_3^2 - 3f_4^2$$
$$- (8f_1 f_2 + 6f_1 f_3 + 4f_1 f_4 + 4f_2 f_3 + 2f_2 f_4)$$
$$+ 27f_1 - 11f_2 - 13f_3 - 3f_4 . \tag{15c}$$

Since these values are the same for all the states belonging to an irreducible representation $[f]$, they are also the average traces of the $SU(4)$ Casimir operators in the subspace labelled by $[f]$.

In Table IX-1, the irreducible representations of $U(N/4)$, the conjugate representations of $SU(4)$, are listed for $k \leq 6$. The invariants that can be

Table IX-1   Irreducible representations and invariant operators
of $U(N/4)$ for $k \leq 6$

Rank	Irreducible representation	Invariant operator
0	[0]	1
1	[1]	$\hat{n}$
2	[2], [11]	$\hat{n}^2$, G_2
3	[3], [21], [111]	$\hat{n}^3$, $\hat{n}G_2$, G_3
4	[4], [31], [22], [211], [1111]	$\hat{n}^4$, $\hat{n}^2G_2$, $\hat{n}G_3$, G_2^2, G_4
5	[5], [41], [32], [311], [221], [2111]	$\hat{n}^5$, $\hat{n}^3G_2$, $\hat{n}^2G_3$, $\hat{n}G_2^2$, $\hat{n}G_4$, G_2G_3
6	[6], [51], [42], [411], [33]	$\hat{n}^6$, $\hat{n}^4G_2$, $\hat{n}^3G_3$, $\hat{n}^2G_2^2$
	[321], [3111], [222], [2211]	$\hat{n}^2G_4$, $\hat{n}G_2G_3$, G_2^3, G_3^2, G_2G_4

constructed out of the Casimir operators for each particle rank are given
in the last column. Since the number of invariants and the number of
irreducible representations are equal, there is no problem to propagate
traces of operators up to maximum particle rank 6.

As an example, we shall work out the centroid and variance of a state
density distribution in $SU(4)$. Since one of the interests here is to examine
how good the symmetry is in nuclear systems, we shall include the zero-
body term in the Hamiltonian usually ignored in other studies. The origin
of the zero-body term comes from the fact that, as mentioned in Chapter
I, we have truncated the Hilbert space for m particles, in part, by ignoring
the particles in the core. In general, the effect of the core on the m-
particle states is included in the one- and two-body terms in the effective
Hamiltonian, and the zero-body term serves only to change the reference
point of the energy scale. However, if the core is polarized (rather than
spherical) it can affect the goodness of a symmetry in the truncated space
and, as a result, the zero-body term must be considered for symmetry
studies in general.

For a centroid, the highest particle rank involved is $k = 2$. Since
there are two irreducible representations in $SU(4)$ for $m = 2$, we need
two defining averages, $\langle \hat{O}(2) \rangle^{2[2]}$ and $\langle \hat{O}(2) \rangle^{2[11]}$, to propagate the average
traces of $\hat{O}(2)$. The trace-equivalent expression of $\hat{O}(2)$ in $SU(4)$ is then

$$\hat{O}_{\text{tr}}(2) = \frac{1}{4}\{\hat{n}(\hat{n} - 5) + G_2(SU(4))\}\, \langle \hat{O}(2) \rangle^{2[2]}$$
$$+ \frac{1}{4}\{\hat{n}(\hat{n} + 3) - G_2(SU(4))\}\, \langle \hat{O}(2) \rangle^{2[11]}.$$

A check of this expression is provided by the fact that in the $m = 2$ space,
the average trace of G_2 from eq. (15a) is 10 for [2] and 6 for [11], and we

find that the averages of $\hat{O}_{\mathrm{tr}}(2)$ reproduce those of $\hat{O}(2)$ in the two-particle space.

To work out the centroid energy, we need to include the contributions of the zero- and one-body terms,

$$\overline{E}_{m[f]} \equiv \langle H \rangle^{m[f]} = \frac{1}{2}(m-1)(m-2)\overline{E}_{0[0]} - m(m-2)\,\overline{E}_{1[1]}$$

$$+ \frac{1}{4}\left\{ m(m-5) + \langle G_2(SU(4)) \rangle^{m[f]} \right\} \overline{E}_{2[2]}$$

$$+ \frac{1}{4}\left\{ m(m+3) - \langle G_2(SU(4)) \rangle^{m[f]} \right\} \overline{E}_{2[11]}, \quad (16)$$

where $\overline{E}_{0[0]}$ defines the trace-equivalent zero-body part of H. The other three input quantities to eq. (16), $\overline{E}_{1[1]}$, $\overline{E}_{2[2]}$, and $\overline{E}_{2[11]}$, are of mixed particle ranks with contributions from $k \leq 2$.

For the variance, operators up to particle rank $k = 4$ are involved,

$$\sigma^2_{m[f]} \equiv \langle H^2 \rangle^{m[f]} - \overline{E}^2_{m[f]}.$$

Using the elementary net, consisting of variances from $SU(4)$ subspaces with $m \leq 4$, this expression (Parikh, 1973) takes the form

$$\sigma^2_{m[f]} = \frac{1}{24}\left\{ (m-1)(m-2)(m-3)(m-4) \right\} \sigma^2_{0[0]}$$

$$+ \frac{m}{6}\left\{ (m-2)(m-3)(m-4) \right\} \sigma^2_{1[1]}$$

$$+ \frac{1}{8}\left\{ m(m-3)(m-4)(m-5) + (m-3)(m-4)\langle G_2 \rangle^{m[f]} \right\} \sigma^2_{2[2]}$$

$$+ \frac{1}{8}\left\{ m(m-3)(m-4)(m+3) - (m-3)(m-4)\langle G_2 \rangle^{m[f]} \right\} \sigma^2_{2[11]}$$

$$- \frac{1}{36}\left\{ m(m-4)(m-5)(m-12) \right.$$

$$\left. + (3m-22)(m-4)\langle G_2 \rangle^{m[f]} + 2(m-4)\langle G_3 \rangle^{m[f]} \right\} \sigma^2_{3[3]}$$

$$- \frac{1}{9}\left\{ m(m-4)(m-5)(m+3) + 8(m-4)\langle G_2 \rangle^{m[f]} \right.$$

$$\left. - (m-4)\langle G_3 \rangle^{m[f]} \right\} \sigma^2_{3[21]}$$

$$- \frac{1}{36}\left\{ m(m-4)(m+3)(m+4) \right.$$

$$\left. - (m-4)(3m+10)\langle G_2 \rangle^{m[f]} + 2(m-4)\langle G_3 \rangle^{m[f]} \right\} \sigma^2_{3[111]}$$

$$+ \frac{1}{576}\left\{ m(m-5)(m-12)(m-21) + 2(3m^2 - 68m + 321)\langle G_2 \rangle^{m[f]} \right.$$

$$\left. + 4(2m-27)\langle G_3 \rangle^{m[f]} + 6\langle G_4 \rangle^{m[f]} + 3\langle G_2^2 \rangle^{m[f]} \right\} \sigma^2_{4[4]}$$

$$+ \frac{1}{64}\left\{ m(m-5)(m-12)(m+3) + 2(m-7)(m+9)\langle G_2 \rangle^{m[f]} \right.$$

$$+ 28\langle G_3 \rangle^{m[f]} - 2\langle G_4 \rangle^{m[f]} - \langle G_2^2 \rangle^{m[f]} \} \, \sigma_{4[31]}^2$$

$$+ \frac{1}{144} \{ m^2(m-5)(m+3) + 8m\langle G_2 \rangle^{m[f]}$$

$$- 4m\langle G_3 \rangle^{m[f]} + 3\langle G_2^2 \rangle^{m[f]} \} \, \sigma_{4[22]}^2$$

$$+ \frac{1}{64} \{ m(m-5)(m+3)(m+4) - 2(m^2 - 6m - 31)\langle G_2 \rangle^{m[f]}$$

$$- 20\langle G_3 \rangle^{m[f]} + 2\langle G_4 \rangle^{m[f]} - \langle G_2^2 \rangle^{m[f]} \} \, \sigma_{4[211]}^2$$

$$+ \frac{1}{576} \{ m(m+3)^2(m+4) - 2(3m+11)(m+3)\langle G_2 \rangle^{m[f]}$$

$$+ 4(2m+9)\langle G_3 \rangle^{m[f]} - 6\langle G_4 \rangle^{m[f]} + 3\langle G_2^2 \rangle^{m[f]} \} \, \sigma_{4[1111]}^2 , \qquad (17)$$

where G_2, G_3, and G_4 are the abbreviations for respectively $G_2(SU(4))$, $G_3(SU(4))$, and $G_4(SU(4))$. This expression can be checked by working out the variances for all the subspaces with $m \leq 4$, using eqs. (15) for the expectation values of the Casimir operators, and arriving at identities. Note that in addition to $\sigma_{0[0]}^2$, the contributions of the zero-body Hamiltonian may also be present in the other terms.

Instead of the elementary net, it would be simpler to make use of particle-hole symmetry to reduce the complexity in preparing the input and use the variances in the spaces with N, $(N-1)$, $(N-2)$, and $(N-3)$ particles to replace the $m = 4$ averages as the defining quantity for the variance. The propagation equation is now somewhat more complicated but can be derived easily from eq. (17).

IX.6   The $SU(4)ST$ scheme

A significant part of the distribution width for a given space symmetry comes from the splitting between different isospins. We can remove this part of the contribution to the distribution width by partitioning the subspace further using the reduction of $SU(4)$ according to S and T,

$$SU(4) \supset SU_S(2) \times SU_T(2). \qquad (18)$$

The decomposition here is more complicated than the subdivision according to isospin for the scalar averaging case given by eq. (7). This is due to the fact that since, in the scalar case, $U(N/2)$ and $U(2)$ representations are conjugates of each other, the isospin T labels the irreducible representations of $U(N/2)$ as well. Here, the conjugation relation exists only between the representations of $U(N/4)$ and $SU(4)$ (or $U(4)$), and the isospin $SU_T(2)$, being further down the chain of group reductions, is no longer related to the space symmetry. Furthermore, as we shall see later, the group structure is complicated by the occurrence of multiplicity for some irreducible

Table IX-2 Irreducible representations and invariant operators
of $SU(4) \supset SU_S(2) \times SU_T(2)$ for $k \leq 4$

Particle rank	Irreducible representation	Casimir invariants	Other invariants
0	$[0](00)$	1	
1	$[1](\frac{1}{2}\frac{1}{2})$	$\hat{n}$	
2	$[2](11)$, $[2](00)$ $[11](10)$, $[11](01)$	$\hat{n}^2$, G_2, S^2, T^2	
3	$[3](\frac{3}{2}\frac{1}{2})$, $[3](\frac{1}{2}\frac{1}{2})$ $[21](\frac{3}{2},\frac{1}{2})$, $[21](\frac{1}{2}\frac{3}{2})$, $[21](\frac{1}{2}\frac{1}{2})$ $[111](\frac{1}{2}\frac{1}{2})$	$\hat{n}^3$, $\hat{n}S^2$, $\hat{n}T^2$ $\hat{n}G_2$, G_3	C_{111}
4	$[4](22)$, $[4](11)$, $[4](00)$ $[31](21)$, $[31](12)$, $[31](11)$, $[31](10)$, $[31](01)$ $[22](20)$, $[22](02)$, $[22](11)$, $[22](00)$ $[211](11)$, $[211](10)$, $[211](01)$ $[1111](00)$	$\hat{n}^4$, T^4, S^4, G_4 $\hat{n}^2S^2$, $\hat{n}^2T^2$ $\hat{n}^2G_2$, $\hat{n}G_3$ G_2^2, G_2T^2 G_2S^2, T^2S^2	$\hat{n}C_{111}$, C_{220} C_{022}, C_{121}

representations, *i.e.*, appearing more than once in the space, at fairly low
particle numbers.

In Table IX-2, the irreducible representations of $SU(4)ST$ up to four
particles are listed together with the available invariant operators of the
groups involved. For $m = 2$, we have four irreducible representations and
there are exactly four Casimir invariants with maximum particle rank 2,
$\hat{n}^2$, G_2, S^2, and T^2 coming respectively from $U(N)$, $SU(4)$, $SU_S(2)$, and
$SU_T(2)$. Since there are a total of six irreducible representations for $m \leq 2$,
we now have a six-term propagation equation for the centroid of a $(0+1+2)$-
body Hamiltonian,

$$\overline{E}_{m[f]ST} = \frac{1}{2}(m-1)(m-2)\ \overline{E}_{0[0]00} - m(m-2)\ \overline{E}_{1[1]\frac{1}{2}\frac{1}{2}}$$

$$- \frac{1}{8}\left\{m - \langle G_2(SU(4))\rangle^{m[f]} + 2S(S+1) + 2T(T+1)\right\}\overline{E}_{2[2]00}$$

$$- \frac{1}{8}\left\{9m - 2m^2 - \langle G_2(SU(4))\rangle^{m[f]} - 2S(S+1) - 2T(T+1)\right\}\overline{E}_{2[2]11}$$

$$+ \frac{1}{8}\left\{3m + m^2 - \langle G_2(SU(4))\rangle^{m[f]} + 2S(S+1) - 2T(T+1)\right\}\overline{E}_{2[11]10}$$

$$+ \frac{1}{8}\left\{3m + m^2 - \langle G_2(SU(4))\rangle^{m[f]} - 2S(S+1) + 2T(T+1)\right\}\overline{E}_{2[11]01},$$

$$(19)$$

arrived at by the same approach as used to obtain eq. (11).

Starting with $k = 3$, we encounter for the first time that there are more defining averages required than the number of independent products of Casimir operators. There are six irreducible representations, but only five linearly independent products of Casimir operators can be formed with maximum particle rank 3. In order to propagate the traces of three-body operators, group invariants other than Casimir operators must be constructed. In general, we can build invariant operators for the chain of reductions given in eq. (18) by considering the product of operators (Quesne, 1976),

$$C_{\alpha\beta\gamma} \equiv (\underbrace{\hat{S} \times \hat{S} \times \cdots \times \hat{S}}_{\alpha} \times \underbrace{\hat{E} \times \hat{E} \times \cdots \times \hat{E}}_{\beta} \times \underbrace{\hat{T} \times \hat{T} \times \cdots \times \hat{T}}_{\gamma})^{(00)},$$

(20)

where $\hat{E} \equiv \hat{\sigma}\hat{\tau}$ is made up of the product of a (2×2) Pauli spin matrix $\hat{\sigma}$ and an isospin matrix $\hat{\tau}$. Only the ST-scalar products are considered here although there is no reason preventing us from constructing non-scalar invariants for propagating more general operators.

Since the operator $C_{\alpha\beta\gamma}$ defined in eq. (20) is a general one, it includes all the Casimir invariants as well. For example,

$$S^2 \sim C_{200}, \qquad T^2 \sim C_{002}, \qquad G_2 \sim C_{020}.$$

On the other hand, the operator $C_{111} = \hat{S} \times \hat{E} \times \hat{T}$ is linearly independent of the five $k = 3$ products of Casimir operators and it supplies the sixth invariant required to propagate three-body operators.

For $k = 4$, sixteen operators are needed. In addition to the product of $\hat{n}$ with C_{111}, we have three new operators C_{220}, C_{022}, and C_{121} that are linearly independent of the $k = 4$ products of Casimir operators. The complete list of sixteen operators is given in Table IX-2.

The process can be continued to higher particle-ranks. For $m = 5$, there are 23 different irreducible representations and only 22 linearly independent products of the invariants can be made from those already given in Table IX-2. The required additional operator is provided by C_{131}. However, starting from $k = 6$ we encounter for the first time the question of multiplicity of irreducible representations. There are altogether 48 different irreducible representations for $m = 6$: 46 of these occur once each, and two, $[42](22)$ and $[321](22)$, occur twice each. The number of invariant scalar operators turns out to be 50. Two *extra* invariant operators are needed to take care of the two (one for $[42](22)$ and one for $[321](22)$) off-diagonal defining averages between two different irreducible representations with the same labels. For symmetries with multiplicities of irreducible representations in the defining spaces, such off-diagonal average traces form a part of the defining quantities of the trace-equivalent operators and are required to propagate the traces.

IX.7 $Sp(2j+1)$ and the seniority scheme

The symplectic symmetry described by the group $Sp(2j+1)$ is of interest in nuclear structure because of its connection with the pairing force which is known to be important near the ground states. The group is composed of those unitary transformations in $U(2j+1)$ that leave the sum of anti-symmetrized bilinear products, $\sum_m (-1)^{m-\frac{1}{2}} \phi_m(i)\phi_{-m}(k)$, invariant. In its elementary form, the symmetry applies only to the space spanned by a single j-orbit.

It is convenient to write the generators of the group in terms of the Racah unit tensors U_{rs}^{Δ}, defined by eq. (C-14). For a single j-orbit, we can omit the subscripts differentiating between orbits and write it in the form

$$U^{\Delta} \equiv [\Delta]^{-\frac{1}{2}} (A \times B)^{\Delta}.$$

From eq. (3) we can obtain the commutation relations among the unit tensors in angular momentum coupled form as

$$[U^r, U^s]^t \equiv (U^r \times U^s)^t - (-1)^{r+s-t}(U^s \times U^r)^t$$
$$= (-1)^{2j+t}(1 - (-1)^{r+s-t})[t]^{\frac{1}{2}} \begin{Bmatrix} r & s & t \\ j & j & j \end{Bmatrix} U^t.$$

The $6j$-symbol enters here because of the need to recouple the angular momenta. For identical particles, the allowed ranks are $\Delta = 0, 1, \ldots, 2j$. Since there are $(2\Delta + 1)$ components for each Δ, the total number of independent unit tensors in the space is

$$\sum_{\Delta=0}^{2j} (2\Delta + 1) = (2j+1)^2.$$

These $(2j+1)^2$ unit tensors form the generators of the $U(2j+1)$ subgroup. Among them, there is a subset of $(j+1)(2j+1)$ odd-rank tensors which constitute the generators of the $Sp(2j+1)$ group. The bilinear Casimir operator,

$$G_2\big(Sp(2j+1)\big) = 2 \sum_{\Delta=\text{odd}} [\Delta](U^{\Delta} \cdot U^{\Delta}), \tag{21}$$

is closely related to the pairing operator (de-Shalit and Talmi, 1963) and, as mentioned earlier, is an important part of the nuclear interaction.

The irreducible representations of $Sp(2j+1)$ are labelled by Young tableaux with two columns of lengths $\frac{1}{2}v \pm t$. The symbol v is known as the seniority, which measures the number of particles which are not coupled in pairs to angular momentum zero, and t, the reduced isospin, is the total isospin formed by coupling the v particles. The irreducible representations of $Sp(2j + 1)$ are therefore distinguished by the number of zero-coupled pairs of nucleons. As a result, it is common practice to use the indices (v, t), instead of the Young tableau, to label the irreducible representations of $Sp(2j + 1)$.

The symplectic group is useful in classifying states in the jj-coupling scheme (Flowers, 1952). Among the generators we have the three components of a rank-one tensor, J_x, J_y, and J_z, which form the generators of the subgroup $R(3)$, the rotation group in three dimensions. Hence, for a single j-orbit with $N = 2(2j + 1)$ single-particle states, we have first the direct product reduction according to isospin given by eq. (7), and then the reduction according to $Sp(2j + 1)$ and $R(3)$ corresponding to the chain

$$U(N/2) \supset Sp(2j + 1) \supset R(3). \tag{22}$$

An m-particle state in this scheme can be represented by $|j^m T(v,t) J x\rangle$, where T, (v,t), and J label respectively the irreducible representations of $U(2)$ (and consequently $U(N/2)$), $Sp(2j + 1)$, and $R(3)$). The additional label x is introduced to resolve any multiplicity of states with the same m, T, (v,t), and J.

For a given j-orbit, there are only three $m = 2$ irreducible representations, $(v,t) = (2,0)$ for the $T = 0$ pair, and $(v,t) = (0,0)$ and $(2,1)$ for the $T = 1$ pair. The trace-equivalent form of a $k = 2$ operator in $Sp(2j + 1)$ is therefore given by the three quantities, $\langle \hat{O}(2) \rangle^{20(0,0)}$, $\langle \hat{O}(2) \rangle^{21(2,0)}$, and $\langle \hat{O}(2) \rangle^{21(2,1)}$. For the invariant operators, we can take $\hat{n}^2$, T^2, and the bilinear Casimir operator G_2 of $Sp(2j + 1)$. Since the expectation value of $G_2(Sp(2j + 1))$ in the irreducible representation $mT(v,t)$ (de-Shalit and Talmi, 1963) is

$$\langle G_2(Sp(2j + 1)) \rangle^{mT(v,t)} = \frac{v}{2}(4j + 8 - v) - 2t(t + 1),$$

the propagation equation for $\hat{O}(2)$ is given by

$$\langle \hat{O}(2) \rangle^{mT(v,t)} = \frac{m(m - 1)}{8} \{3 \langle \hat{O}(2) \rangle^{21(2,1)} + \langle \hat{O}(2) \rangle^{20(2,0)}\}$$

$$+ \frac{4T(T + 1) - 3m}{8(2j + 1)} \{(2j + 3) \langle \hat{O}(2) \rangle^{21(2,1)}$$

$$- (2j + 1) \langle \hat{O}(2) \rangle^{20(2,0)} - 2 \langle \hat{O}(2) \rangle^{21(0,0)}\}$$

$$+ \frac{1}{2j + 1} \left\{ \frac{1}{4}(m - v)(4j + 5 - m - v) + t(t + 1) - \frac{3}{4}v \right\}$$

$$\times \{ \langle \hat{O}(2) \rangle^{21(0,0)} - \langle \hat{O}(2) \rangle^{21(2,1)} \}. \tag{23}$$

To obtain the centroid of a complete (0+1+2)-body Hamiltonian, we add to eq. (23) the contributions from the zero- and one-body parts. Since we are dealing with a single j-orbit here, the one-body contribution is simply the single-particle energy ϵ_j of the orbit. Consequently, the centroid energy can be written as

$$\overline{E}_{mT(v,t)} = \langle H(2) \rangle^{mT(v,t)} + \overline{E}_{00(0,0)} + m\epsilon_j \,,$$

where $\overline{E}_{00(0,0)}$ comes purely from the zero-body part of the Hamiltonian and $\langle H(2) \rangle^{mT(v,t)}$ is given by eq. (22).

The propagation of the variance is complicated by the fact that the number of irreducible representations of $Sp(2j+1)$ increases very fast with particle number even for orbits with moderately large j values. For example, there are seven different (v,t) representations for $m = 4$ in the $j = 5/2$ orbit, and nine in the $j = 7/2$ orbit. Except for identical particle spaces, it is not a simple matter to obtain the variance by propagation (Parikh, 1978).

IX.8   *SU*(3) symmetry

In the reduction of $U(N)$ to $U(N/4) \times U(4)$, we can decompose the irreducible representations of $U(N/4)$ further according to the chain

$$U(N/4) \supset SU(3) \supset R(3) \,. \tag{24}$$

Since the total orbital angular momentum L labels the irreducible representations of $R(3)$ here, $SU(3)$ is useful in the classification of states in the LS-coupling scheme and plays the same role as $Sp(2j+1)$ in the jj-coupling scheme. $SU(3)$ symmetry is also interesting because of its connection with rotational bands in nuclei (Elliot, 1958). In its elementary form, the $SU(3)$ scheme applies only to a space made up of single-particle states belonging to the same major shell.

Young tableaux describing the irreducible representations of $SU(3)$ have three rows of lengths f_1, f_2, and f_3,

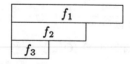

with the conditions that $f_1 \geq f_2 \geq f_3$ and $\sum_i f_i = n_q m$, where n_q is the number of oscillator quanta associated with a single particle in the major shell. For the 1p-shell, $n_q = 1$; the ds-shell, $n_q = 2$; the fp-shell, $n_q = 3$; and so on. Since representations given by Young tableaux differing only in the number of complete columns are indistinguishable in $SU(N)$, only two indices are needed to label the irreducible representations of $SU(3)$. Usually

$$\lambda = f_1 - f_2 \qquad \text{and} \qquad \mu = f_2 - f_3$$

are taken as the standard notation in nuclear physics.

A state in the $SU(3)LS$ scheme can be written as $|m[f](\lambda\mu)LSTx\rangle$, where $[f]$, $(\lambda\mu)$, and L label respectively $SU(N/4)$, $SU(3)$, and $R(3)$ whereas T and S label the spin and isospin in the reduction of $SU(4)$. The label x is again used to resolve any multiplicity in the representation.

The eight generators of $SU(3)$ are commonly taken to be the three components of $\hat{L}$ and the five components of the quadrupole operator $\hat{Q}$. There are two Casimir invariants, a bilinear one,

$$G_2(SU(3)) = \frac{1}{4}\{3\hat{L}\cdot\hat{L} + \hat{Q}\cdot\hat{Q}\}, \qquad (25)$$

having the expectation value

$$\langle G_2(SU(3))\rangle^{(\lambda\mu)} = \{(\lambda+\mu)(\lambda+\mu+3) - \lambda\mu\}, \qquad (26)$$

in a state of definite $(\lambda\mu)$, and a cubic one (Biedenharn, 1962) which we shall not use here.

Table IX-3   The irreducible representations of $U(6) \supset SU(3)$
in the ds-shell for $m \leq 4$

m	$[f]$	$(\lambda\mu)$
0	[0]	(00)
1	[1]	(20)
2	[2]	(40), (02)
	[11]	(21)
3	[3]	(60), (22), (00)
	[21]	(41), (22), (11)
	[111]	(30), (03)
4	[4]	(80), (42), (04), (20)
	[31]	(61), (42), (23), (31), (12), (20)
	[22]	(42), (31), (04), (20)
	[211]	(50), (23), (31), (12), (01)
	[1111]	(12)

In the ds-shell, a single-particle state belongs to the $SU(3)$ irreducible representation [1](20), and in the fp-shell to [1](60). For $m = 2$, there are three irreducible representations in the ds-shell (see Table IX-3) and four in the fp-shell, $[2](00)$, $[2](22)$, $[11](41)$, and $[11](03)$. With the three Casimir invariants of maximum particle rank 3, $\hat{n}^2$, $G_2(U(6))$, and $G_2(SU(3))$ in the $U(24) \supset U(6) \supset SU(3)$ reduction chain, we can propagate the average of any two-body, angular momentum scalar operators in the ds-shell. To obtain the centroid energy of a $(0+1+2)$-body Hamiltonian, we can include $\hat{n}$ for the one-body part and a constant for the zero-body part. The result, given in Parikh (1972), is

$$
\begin{aligned}
\overline{E}_{m[f](\lambda\mu)} = \ & \frac{1}{2}\{m^2 - 3m + 2\}\overline{E}_{0[0](00)} - \{m^2 - 2m\}\overline{E}_{1[1](20)} \\
& + \frac{1}{36}\{7m^2 - 15m - 3\langle G_2(U(6))\rangle^{m[f]} + 2\langle G_2(SU(3))\rangle^{(\lambda\mu)}\}\overline{E}_{2[2](40)} \\
& + \frac{1}{18}\{m^2 + 21m - 3\langle G_2(U(6))\rangle^{m[f]} - \langle G_2(SU(3))\rangle^{(\lambda\mu)}\}\overline{E}_{2[2](02)} \\
& + \frac{1}{4}\{m^2 - 5m + \langle G_2(U(6))\rangle^{m[f]}\}\overline{E}_{2[11](21)},
\end{aligned}
\tag{27}
$$

where the average traces $\langle G_2(U(6))\rangle^{m[f]}$ can be obtained from the average traces of $G_2(SU(4))$ in the conjugate representation $[\tilde{f}]$ using eq. (15a).

Again we have difficulty in propagating the variance. As shown in Table IX-3, the number of irreducible representations is already seven for $m = 3$ in the ds-shell, and for $m = 4$ the number is 21. It does not seem possible to find enough invariant operators in the limited algebra for propagating any operator beyond the centroid in $SU(3)$.

Several other group symmetries are also of interest. For identical particles, propagation of widths in $Sp(2j + 1)$ is possible only up to $j = 9/2$. The question of extending the method to include several j-orbits has been examined by Quesne and Spitz (1974) using the quasi-spin idea. Averaging techniques have also been used to study $Sp(3, R)$ symmetry for collective motions in nuclei (Draayer and Rosensteel, 1983). It is likely that more progress will be made in these directions.

The general question of the conditions under which one-step propagation is possible has been examined by Quesne (1975). The results seem to indicate that one must consider the alternatives of iterative and other types of propagation in order that averaging methods can be applied to study symmetries other than those described above.

Chapter X

Applications

In this chapter selected applications to problems in nuclear physics are given as illustrations. It is by no means a complete survey; the aim here is to show how some of the methods are carried out in practice and to exhibit the types of problem that statistical spectroscopy is trying to address.

The examples can be classified into three broad groups. In the earlier days of the development, the primary concern in statistical spectroscopy was to establish the method. Consequently, the tendency was to emphasize comparisons with exact calculations of standard nuclear spectroscopic quantities such as energy level positions and transition rates in the ground state region. These applications form the first group. However they are not the most interesting examples for our purpose here since statistical spectroscopy is not meant to compete in these areas. In section 1, a binding energy study is used to give a flavour of the comparison with shell model results.

A second group, consisting of applications such as state density and transition strength distributions, is more closely related to a statistical approach. Statistical spectroscopic methods are in principle ideal for such studies. The examples in sections 2 to 5 are chosen to show the ease with which such studies can be carried out and the advantages in using such an approach.

The third category, given in sections 6 to 10, consists of the more novel studies that are unique to statistical spectroscopy. Concepts such as the correlation between different Hamiltonians, and between a Hamiltonian and an excitation operator, are not commonly used in other nuclear structure studies. These examples are chosen to indicate the possible new insights that may be obtained.

X.1 Binding energy

In statistical spectroscopy, only the variation of a physical quantity as a function of energy is calculated. To compare with the results of detailed spectroscopy in the low-lying regions, it is often necessary to convert the continuous function obtained using a set of low-order moments into discrete values associated with individual states. The procedure for making such a conversion is demonstrated by the binding energy study.

In many respects, the binding energy of a nucleus is one of the least favourable quantities to apply statistical spectroscopic methods. First, it is measured at the ground state, the state that is the furthest away from the centroid of an eigenvalue distribution; the intrinsic accuracy of the method is therefore the worst here. Second, statistical spectroscopy can only give the *most probable* location of a state: the actual position can deviate from the most probable position by one local level spacing on the average because of fluctuations, as observed in random matrix studies. Because of the exponential increase in level density with excitation energy, the local level spacing in energy units is a maximum at the ground state. Consequently, an *error* of the order of one local spacing unit, although quite acceptable in statistical spectroscopy, is often too large in energy units for many purposes. In spite of these problems, the binding energy test came out rather favourably for statistical spectroscopy.

Since the ground state of a nucleus has definite spin and isospin, we should in principle use the fixed-JT averaging method to obtain the energy, as was done by Lougheed and Wong (1975). However, for illustrative purposes, we shall use the configuration-isopin averaging method here; the simplicity of the mT method outweighs the improvements in the accuracy obtained by fixed-JT averaging. The separations between state densities of different isospin obtained by mT-averaging is very useful since the lowest states of different isospins are well separated in energy.

In mT-averaging, the intensity of the strength distribution at energy E in a space with particle number m and isospin T is given by the sum over all mT-subspace distributions,

$$I_{mT}(E) = \sum_m d_{mT}\rho_{mT}(E). \tag{1}$$

Each distribution $\rho_{mT}(E)$ is specified by a set of moments $\langle H^r \rangle^{(mT)}$ in the mT-subspace. If the ground state spin-isospin is (J_0, T_0), we can take the most probable location of the ground state to be the energy E_0 satisfying the relation

$$\int_{-\infty}^{E_0} I_{mT_0}(x)\, dx = \frac{1}{2}(2J_0 + 1)(2T_0 + 1). \tag{2}$$

The rationale behind this *recipe* is illustrated by Fig. X-1.

Given a smooth, continuous distribution $I_{mT}(E)$, we wish to determine the most probable position of a discrete state. Let us view this problem first from the opposite direction. If we are given a set of eigenvalues $\{E_i\}$ for $i = 1, 2, \ldots, d$, say, from a shell model calculation, we can construct a graph showing the distribution of eigenvalues by plotting the integrated

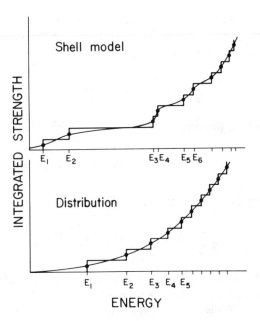

Fig. X-1. Integrated strength of a discrete spectrum and its smooth counterpart. Because of fluctuations, the low-lying shell model levels do not have the same behaviour as in the distribution model.

strength as a function of energy,

$$F_T(E) = \int_{-\infty}^{E} I_{mT}(x)\, dx. \tag{3}$$

For a discrete spectrum of isospin T, the plot is a step function with $F_T(E)$ increasing by $(2J_i + 1)(2T + 1)$ at the location of an eigenstate with spin-isospin (J_i, T), as shown in Fig. X-1. In the middle of the spectrum, the state density is sufficiently high and the step function is indistinguishable from a continuous function. On the other hand, in the ground state region (and the high energy end as well for the partial result in a finite space), the density is low and the discrete steps must be removed somehow, for example, by drawing a smooth curve through the midpoints of the steps as shown in Fig. X-1. The smooth curve then has the property that

$$F(E_k) = \sum_{i}^{k-1}(2J_i + 1)(2T + 1) + \frac{1}{2}(2J_k + 1)(2T + 1), \tag{4}$$

at E_k, the energy of level k.

In statistical spectroscopy we start with a smooth curve. Because of fluctuations, it is not possible in general to reproduce the distribution exactly by using only the low-order moments, but we assume that we can come fairly close to it. Since the ground state is the lowest member of the spectrum, we obtain eq. (2) from eqs. (3) and (4).

The input to our binding energy calculation is a set of configuration-isospin moments. In the Gaussian approximation, each mT-distribution $\rho_{mT}(E)$ is completely given by the centroid C_{mT} and width σ_{mT}. Since

$$\frac{1}{\sigma\sqrt{2\pi}} \int_{-\infty}^{y} \exp{-\frac{(x-c)^2}{2\sigma^2}} \, dx = \frac{1}{2}\left\{1 + \mathrm{erf}\left(\frac{y-c}{\sigma\sqrt{2}}\right)\right\},$$

where the error function is defined by

$$\mathrm{erf}(z) = \frac{2}{\sqrt{\pi}} \int_{0}^{z} e^{-t^2} \, dt,$$

eq. (1) reduces to

$$\frac{1}{2}\sum_{m} d_{mT_0}\left\{1 + \mathrm{erf}\left(\frac{E_0 - C_{mT_0}}{\sqrt{2}\,\sigma_{mT_0}}\right)\right\} = \frac{1}{2}(2J_0 + 1)(2T_0 + 1). \qquad (5)$$

A simple search procedure may be used to find the value of E_0 satisfying eq. (5) since there is no need to obtain E_0 to high accuracy.

Table X-1 lists the binding energies with respect to an ^{16}O core for ds-shell nuclei with $A = 20$ to 36. The Coulomb energy, E_{Coul}, has been removed from the experimental value of each nucleus by Chung (1976) using the positions of isobaric analogy states and other relevant information. Both the shell model and (mT)-averaging results were obtained with the *particle* interaction of Wildenthal and Chung (1979) for $A \leq 28$ and the *hole* interaction of the same authors for $A > 28$. The 63 two-body matrix elements and the three single-particle energies that define the ds-Hamiltonian were obtained by least-square fitting to experimental data, including the binding energies, of the appropriate nuclei.

From Table X-1, one can see that the shell model results agree very well with the experimental values with a root-mean-square (rms) error of 0.7 MeV. This is perhaps not surprising since binding energies form an important part of the input data to the least-square fitting procedure used to obtain the interactions. The good agreement demonstrates that the interactions used here are realistic ones. Furthermore, the shell model results are obtained by diagonalizing the complete ds-shell Hamiltonian matrices and therefore yield the exact values for the interaction in the space. It is therefore more useful to compare the statistical spectroscopy results with those given by the shell model since our concern here lies more in the errors introduced by the approximations in mT-averaging. For the 53 cases in the table, a rms value of 4.7 MeV is obtained for the differences between the mT-averaging and shell model results. For binding energies, the error is rather a large one; on the other hand, the mT-averaging calculations are far simpler, by several orders of magnitude in terms of computer time, for nuclei in the middle of the ds-shell.

Table X-1 Binding energy of ds-shell nuclei with respect to ^{16}O core
(Corrected for Coulomb energy contribution by Chung, 1976)

A	2J	2T	E_{exp}	E_{sm}	E_{mT}	A	2J	2T	E_{exp}	E_{sm}	E_{mT}
^{20}O	0	4	−23.75	−23.98	−24.4	^{28}Mg	0	4	−120.48	−121.44	−129.1
^{21}O	5	5	−26.30	−27.69	−25.1	^{29}Mg	3	5	−124.12	−123.78	−126.6
^{22}O	0	6	−32.20	−34.78	−33.7	^{26}Al	10	0	−105.80	−106.45	−114.9
^{20}F	4	2	−30.22	−30.47	−31.9	^{30}Al	6	4	−141.43	−141.01	−145.1
^{21}F	5	3	−38.39	−38.51	−39.4	^{31}Al	5	5	−148.62	−149.14	−150.2
^{22}F	8	4	−43.56	−43.61	−45.2	^{28}Si	0	0	−135.70	−138.03	−148.1
^{23}F	5	5	−51.30	−51.30	−51.7	^{29}Si	1	1	−144.18	−143.46	−149.9
^{20}Ne	0	0	−40.48	−40.60	−40.2	^{30}Si	0	2	−154.79	−154.59	−160.4
^{21}Ne	3	1	−47.24	−47.29	−48.6	^{31}Si	3	3	−161.37	−161.09	−164.4
^{22}Ne	0	2	−57.61	−57.64	−61.3	^{32}Si	0	4	−170.59	−170.80	−173.3
^{23}Ne	5	3	−62.80	−62.88	−65.7	^{33}Si	3	5	−175.14	−175.29	−174.4
^{24}Ne	0	4	−71.67	−72.04	−77.2	^{30}P	2	0	−155.46	−155.09	−163.4
^{25}Ne	1	5	−75.97	−75.95	−80.8	^{31}P	1	1	−167.77	−167.88	−174.2
^{22}Na	6	0	−58.23	−58.23	−62.6	^{32}P	2	2	−175.64	−175.74	−181.4
^{23}Na	3	1	−70.76	−70.76	−75.2	^{33}P	1	3	−185.78	−186.34	−189.1
^{24}Na	8	2	−77.65	−77.75	−82.9	^{34}P	2	4	−192.02	−192.09	−193.6
^{25}Na	5	3	−86.66	−87.04	−92.3	^{35}P	1	5	−200.48	−200.96	−198.1
^{26}Na	6	4	−92.28	−92.50	−98.3	^{32}S	0	0	−182.64	−182.85	−189.3
^{27}Na	3	5	−99.07	−99.43	−104.4	^{33}S	3	1	−191.28	−191.32	−195.3
^{28}Na	2	6	−102.66	−102.76	−107.9	^{34}S	0	2	−202.70	−203.08	−205.9
^{29}Na	5	7	−106.94	−107.50	−107.4	^{35}S	3	3	−209.69	−209.58	−209.7
^{30}Na	2	8	−109.30	−110.01	−110.7	^{36}S	0	4	−219.58	−219.50	−217.7
^{31}Na	5	9	−115.14	−113.62	−108.6	^{34}Cl	6	0	−202.55	−202.63	−206.9
^{24}Mg	0	0	−87.11	−87.49	−94.3	^{35}Cl	3	1	−215.34	−215.69	−217.1
^{25}Mg	5	1	−94.44	−94.83	−101.1	^{36}Cl	4	2	−223.82	−223.84	−222.4
^{26}Mg	0	2	−105.51	−106.33	−115.6	^{36}Ar	0	0	−230.43	−230.75	−231.8
^{27}Mg	1	3	−111.97	−109.99	−121.4						

The accuracy can be improved somewhat by deducing, rather than the ground state itself, the energy of a reference state k among the low-lying excited states. From eqs. (3) and (4), we have

$$\int_{-\infty}^{E_k} I_{mT}(x)\, dx = \sum_i^{k-1} (2J_i + 1)(2T + 1) + \frac{1}{2}(2J_k + 1)(2T + 1). \quad (6)$$

If the excitation energy E_x of the reference state is known, we can obtain the ground state energy by taking the difference

$$E_0 = E_k - E_x.$$

In this way, the rms error is reduced to less than 3 MeV for the 36 of the 53 cases for which excited states are available to serve as the reference states.

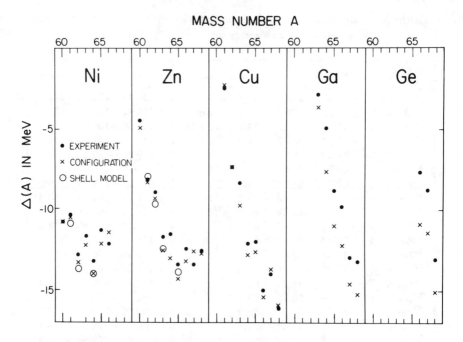

Fig. X-2. Difference between calculated and experimental binding energies of some fp-shell nuclei in terms of $\Delta(A)$.

Chang, French, and Thio (1971) have calculated the binding energies of some fp-shell nuclei assuming an inert ^{56}Ni core. The interaction used is based on the Hamada-Johnson force for nucleon-nucleon scattering and renormalized by Kuo (1967) for limiting the active space to the fp-shell. In order to reduce a systematic deviation from the experimental values, Chang *et al.* added a small correction term, $0.07\binom{m}{2} + 0.11T(T + 1)$, to the eigenvalues obtained with the effective Hamiltonian for a state with m nucleons in the fp-space and isospin T. Since the dimensions of the spaces involved here are much larger than those in the ds-shell cases, only a few shell model results, mostly for the nickel and copper isotopes, are available. Where the comparison can be made, the agreement with experimental values is within 0.5 MeV, indicating that the interaction used is a realistic one.

In Fig. X-2, the distribution results are compared with the experimental values and the available shell model results in terms of the quantity

$$\Delta(A) = E_0 - E_{\mathrm{Coul}} + 8.07 \times (N - 28) + 7.29 \times (Z - 28),$$

the mass excess in MeV of nucleons outside a ^{56}Ni core, obtained by removing the contributions due to the neutrons (8.07 MeV each) and protons

(7.29 MeV each) from the difference in mass excess between the nucleus and the ^{56}Ni core. The values of E_{Coul} are obtained from Coulomb displacement energies and are in the range -8.9 to -9.5 MeV for $Z = 29$, -18.6 to -19.3 MeV for $Z = 30$, -28.3 to -29.2 for $Z = 31$, and -38.1 to -39.0 MeV for $Z = 32$ nuclei. In most cases, the distribution results are obtained using eq. (6) rather than eq. (5). The agreement appears to be better than for the ds-shell nuclei, but the two comparisons are not quite on the same footing. However, in both cases we find that the accuracies of the distribution results, though not adequate to be used in most studies involving binding energies, are better than expected *a priori* for statistical spectroscopy.

X.2   State density and nuclear partition function

It is well known that the state density $\omega_A(E)$, the number of states per unit energy for a nucleus made up of A nucleons, increases roughly exponentially with the square root of the excitation energy E. Bethe (1937) derived the relation

$$\omega_A(E) = \frac{1}{12a^{1/4}E^{5/4}} \exp\{2\sqrt{aE}\}, \tag{7}$$

using statistical mechanics arguments. This expression is often referred to as the Fermi gas model since the nucleons inside a nucleus are treated essentially as non-interacting Fermi particles.

A brief review of the derivation is useful as background for the statistical spectroscopy approach to this problem. The Hamiltonian used in the derivation of eq. (7) is taken to be purely one-body and is given by a set of single-particle energies $\{\epsilon_i\}$. This is one of the major assumptions made in the derivation: the ignored two-body part of the Hamiltonian is important since it depresses the ground state energy from which the excitation energy E in eq. (7) is measured. We shall return to this point later.

For a one-body Hamiltonian, the number of states per unit energy interval at energy E is given by

$$\omega_A(E) = \sum_m \sum_m \delta(A - m)\,\delta(E - E_m), \tag{8}$$

where

$$m = \sum_i m_i, \qquad E_m = \sum_i m_i \epsilon_i. \tag{9}$$

Each single-particle orbit here consists of only one state so that m_i is either 0, if the state is unoccupied, or 1 if it is occupied. As usual, the bold face letter m stands for the set of occupancies $(m_1, m_2, \cdots)$.

Instead of calculating $\omega_A(E)$ directly, it is easier to derive eq. (7) for $\omega_A(E)$ using the partition function

$$Z(\alpha,\beta) = \int_0^\infty \int_0^\infty \omega_A(E) \exp\{\alpha A - \beta E\}\, dA\, dE, \qquad (10)$$

the Laplace transform of $\omega_A(E)$. The parameters α and β correspond respectively to the chemical potential μ and temperature T in statistical mechanics. For a system satisfying eqs. (8) and (9), the partition function can be simplified to

$$Z(\alpha,\beta) = \sum_{mm} \exp\{\alpha m - \beta E_m\}$$
$$= \sum_{mm} \exp\Big\{\sum_i m_i(\alpha - \beta\epsilon_i)\Big\} = \prod_i \{1 + \exp(\alpha - \beta\epsilon_i)\}, \quad (11)$$

where the second equality is obtained from eq. (9), and the final result using the fact that $m_i = 0$ or 1.

The second assumption made to derive eq. (7) is that the single-particle spectrum,

$$g(\epsilon) = \sum_i \delta(\epsilon - \epsilon_i),$$

can be approximated by a continuous distribution. This is true if the single-particle states are closely spaced. In practice, this assumption does not seem to affect the state density for large A.

The product over i in eq. (11) can be changed into a sum by taking the logarithm of $Z(\alpha,\beta)$, and thence to an integral by taking the single-particle spectrum $g(\epsilon)$ to be a continuous one,

$$\ln Z(\alpha,\beta) = \sum_i \ln\{1 + \exp(\alpha - \beta\epsilon_i)\}$$
$$\longrightarrow \int_0^\infty g(\epsilon) \ln\{1 + \exp(\alpha - \beta\epsilon)\}\, d\epsilon. \qquad (12)$$

The zeropoint of the energy scale of the single-particle spectrum is set in such a way that

$$g(\epsilon) = 0 \qquad \text{for} \qquad \epsilon < 0.$$

Since the logarithmic factor in eq. (12) goes to zero very quickly for $\epsilon > \alpha/\beta$, the integral can be expanded in a series involving the derivatives of $g(\epsilon)$ at $\epsilon = \alpha/\beta$ (Bohr and Mottelson, 1969, p. 282),

$$\ln Z(\alpha,\beta) = \int_0^{\alpha/\beta} g(\epsilon)(\alpha - \beta\epsilon)\, d\epsilon + \frac{\pi^2}{6\beta} g(\alpha/\beta) + \frac{7\pi^4}{360\beta^3} g''(\alpha/\beta) + \cdots,$$

where the double prime indicates the second derivative of g. The higher-order terms, not shown, involve the higher-order derivatives.

We now return to the state density by applying an inverse Laplace transform to the partition function,

$$w_A(E) = \left(\frac{1}{2\pi i}\right)^2 \int \int_{-i\infty}^{+i\infty} Z(\alpha, \beta) \exp\{-\alpha A + \beta E\} \, d\alpha \, d\beta. \qquad (13)$$

The integral can be carried out by the saddle-point method since the main contribution comes only from a small region around (α_0, β_0) given by the conditions

$$\begin{cases} \frac{\partial \ln Z}{\partial \alpha} - A = 0, \\ \frac{\partial \ln Z}{\partial \beta} + E = 0. \end{cases} \qquad (14)$$

This constitutes the third approximation in the derivation of the Fermi gas form, and with this approximation we obtain the result

$$w_A(E) = \frac{Z(\alpha_0, \beta_0)}{2\pi |D|^{\frac{1}{2}}} \exp\{-\alpha_0 A + \beta_0 E\}. \qquad (15)$$

The determinant,

$$D = \begin{vmatrix} \frac{\partial^2 \ln Z}{\partial \alpha^2} & \frac{\partial^2 \ln Z}{\partial \alpha \, \partial \beta} \\ \frac{\partial^2 \ln Z}{\partial \beta \, \partial \alpha} & \frac{\partial^2 \ln Z}{\partial \beta^2} \end{vmatrix}_{\alpha=\alpha_0, \beta=\beta_0}$$

is determined by the second derivatives of $\ln Z$ evaluated at (α_0, β_0).

For a one-body Hamiltonian, the ground state of the A-particle system is formed by filling all the single-particle levels up to the Fermi energy ϵ_F. As a result, we have

$$A = \int_0^{\epsilon_F} g(\epsilon) \, d\epsilon, \qquad E_0 = \int_0^{\epsilon_F} \epsilon g(\epsilon) \, d\epsilon. \qquad (16)$$

We now make the fourth assumption, viz., that the single-particle spectrum $g(\epsilon)$ is sufficiently constant that we may ignore all the derivatives of $g(\epsilon)$. At the saddle point, we then have

$$A = \int_0^{\alpha_0/\beta_0} g(\epsilon) \, d\epsilon, \qquad E = \int_0^{\alpha_0/\beta_0} \epsilon g(\epsilon) \, d\epsilon + \frac{\pi^2}{6\beta_0^2} g(\alpha_0/\beta_0). \qquad (17)$$

Comparing eqs. (16) with (17), we can make the identifications

$$\epsilon_F = \alpha_0/\beta_0, \qquad E_x = E - E_0 = +\frac{\pi^2}{6\beta_0^2} g(\epsilon_F), \qquad (18)$$

and eq. (15) simplifies to

$$\omega_A(E_x) = \frac{1}{E\sqrt{48}} \exp\left\{2\sqrt{\frac{\pi^2}{6} g(\epsilon_F) E_x}\right\}. \tag{19}$$

This is the state density for systems of identical particles.

To obtain eq. (7), we need to consider protons and neutrons with single-particle spectra $g_p(\epsilon)$ and $g_n(\epsilon)$, respectively. The complete state density is a convolution of the contributions from both types of particles. Eq. (7) is obtained by redefining the zero of the energy scale to be at E_0, so that $E_x \to E$, and using

$$a = \frac{\pi^2}{6}\left\{g_p(\epsilon_{pF}) + g_n(\epsilon_{nF})\right\},$$

where ϵ_{pF} and ϵ_{nF} are the Fermi energies of the protons and neutrons respectively.

Among the four approximations made in deriving the Fermi gas result, the one-body Hamiltonian and the constant single-particle spectrum are the most serious ones. We shall now show how statistical spectroscopy can remedy both of these problems.

First we shall try to incorporate a realistic single-particle spectrum into the calculation of $\omega_A(E)$. Traditionally a combinatorial approach is used (Hillman and Grover, 1969). Conceptually this is a simple approach and is similar to the idea of configuration averaging. Since the Hamiltonian is still purely one-body, the energy of a state is determined by the occupancies of the single-particle orbits

$$E_m = \sum_r m_r \epsilon_r.$$

The degeneracy of states at energy E_m is d_m, the number of states in configuration m. In the absence of two-body interactions, the configuration strength distribution is a delta function, and the state density can be obtained by averaging over an energy interval ΔE,

$$\omega_A(E) = \frac{1}{\Delta E} \sum_m{}' d_m, \tag{20}$$

where the prime indicates that the summation is restricted to those configurations with E_m in the region $E \pm \Delta E/2$. The method is laborious at the higher energies because of the large number of contributing single-particle orbits. (For values near the ground state, the effect of the two-body interaction is important and the combinatorial result is not useful since it does not contain the two-body effects.) For example, Hillman and Grover (1969) used a total of 62 orbits, half for neutrons and half for protons. In such a large space, the number of possible configurations is huge and an exact calculation, even with a one-body Hamiltonian, is quite time consuming. Since in eq. (20) we are averaging over many configurations, it seems rather superfluous to calculate the distribution in each configuration separately and exactly.

Using statistical spectroscopy methods, we can reduce the number of configurations drastically by grouping orbits into *shells* and using the distribution of nucleons in different shells, rather than in different single-particle orbits, as the basic subspace. In each shell the strength distribution is no longer a delta function, since in general a shell consists of several nondegenerate single-particle orbits. For a *traceless* one-body Hamiltonian,

$$H(1) = \sum_r (\epsilon_r - \bar{\epsilon})\hat{n}_r, \qquad \bar{\epsilon} = \frac{1}{N}\sum_r N_r \epsilon_r, \qquad N = \sum_r N_r,$$

the moments $M_\mu(m)$ of a strength distribution for m particles in the shell can be worked out easily, and the lowest eight orders are given by

$$M_1(m)= m\bar{\epsilon},$$
$$M_2(m)= P(2,1)\Lambda_2,$$
$$M_3(m)= [P(3,1) - P(3,2)]\Lambda_3,$$
$$M_4(m)= 3P(4,2)\Lambda_2^2 + [P(4,1) - 4P(4,2) + P(4,3)]\Lambda_4,$$
$$M_5(m)= 10[P(5,2) - P(5,3)]\Lambda_2\Lambda_3$$
$$\qquad +[P(5,1) - 11P(5,2) + 11P(5,3) - P(5,4)]\Lambda_5,$$
$$M_6(m)= 15P(6,3)\Lambda_2^3 + [P(6,2) - 4P(6,3) + P(6,4)]15\Lambda_2\Lambda_4$$
$$\qquad +[P(6,2) - 2P(6,3) + P(6,4)]10\Lambda_3^2$$
$$\qquad +[P(6,1) - 26P(6,2) + 66P(6,3) - 26P(6,4) + P(6,5)]\Lambda_6,$$
$$M_7(m)= [P(7,3) - P(7,4)]105\Lambda_2^2\Lambda_3$$
$$\qquad +[21P(7,2) - 231P(7,3) + 231P(7,4) - 21P(7,5)]\Lambda_2\Lambda_5$$
$$\qquad +[35P(7,2) - 175P(7,3) + 175P(7,4) - 35P(7,5)]\Lambda_3\Lambda_4$$
$$\qquad +[P(7,1) - 57P(7,2) + 302P(7,3) - 302P(7,4)$$
$$\qquad\qquad +57P(7,5) - P(7,6)]\Lambda_7,$$
$$M_8(m)= [P(8,1) - 120P(8,2) + 1191P(8,3) - 2416P(8,4) + 1191P(8,5)$$
$$\qquad\qquad -120P(8,6) + P(8,7)]\Lambda_8$$
$$\qquad +[35P(8,2) - 280P(8,3) + 630P(8,4) - 280P(8,5) + 35P(8,6)]\Lambda_4^2$$
$$\qquad +[56P(8,2) - 672P(8,3) + 1232P(8,4) - 672P(8,5) + 56P(8,6)]\Lambda_3\Lambda_5$$
$$\qquad +[28P(8,2) - 728P(8,3) + 1848P(8,4) - 728P(8,5) + 28P(8,6)]\Lambda_2\Lambda_6$$
$$\qquad +[280P(8,3) - 56P(8,4) + 280P(8,5)]\Lambda_2\Lambda_3^2$$
$$\qquad +[210P(8,3) - 840P(8,4) + 210P(8,5)]\Lambda_2^2\Lambda_4 + 105P(8,4)\Lambda_2^4, \qquad (21)$$

where we have put

$$\Lambda_q = \sum_r N_r(\epsilon_r - \bar{\epsilon})^q, \qquad \text{and} \qquad P(p,t) = \binom{N}{m}^{-1}\binom{N-p}{m-t},$$

to simplify the notation.

For a configuration consisting of m_r particles in shell r, the moments of the strength distribution can be expressed in terms of the scalar moments of each shell by a multinomial expansion,

$$M_\mu(m) = \langle H(1)^\mu \rangle^{(m)} = \left(\sum_r m_r \epsilon_r \right)^\mu = \mu! \sum_{\{\mu_r\}} \prod_r \frac{M_{\mu_r}(m_r)}{\mu_r!},$$

where the sum runs over all the possible partitions of μ into non-negative integers $(\mu_1, \mu_2, \ldots)$, and $\sum_r \mu_r = \mu$. The product is taken over all the shells in the space.

In an application to ^{56}Fe (Haq and Wong, 1980), the space originally used by Hillman and Grover is divided into sixteen shells, half for neutrons and half for protons. The results are shown in Fig. X-3. Compared with the exact combinatorial values (histogram), the statistical spectroscopy results (solid line) are found to be very accurate. The reduction in computing time achieved by combining orbits into shells is about a factor of twenty up to the energies shown in the figure. Since the number of participating configurations increases roughly exponentially with the square root of the energy, the savings are expected to be even more substantial at higher excitation energies. The results of the Fermi-gas model, eq. (7), are shown for $a = 7.2 \, \text{MeV}^{-1}$, a value determined by fitting the combinatorial results. The model is known to be inadequate except in narrow energy regions: this can also be seen in the figure from the fact that it is impossible to obtain the correct energy dependence compared with the combinatorial results over the whole region. Also shown in the figure are the results of using a smooth (but not constant) form for $g(\epsilon)$. As pointed out by Kahn and Rosenzweig (1969), only the low-energy part of the state density is significantly affected by ignoring the fluctuations in the single-particle density.

The agreement with experimental data for eq. (7) can be improved by shifting the energy scale by an amount Δ. Empirically, the back-shifted Fermi-gas model formula,

$$\omega_A(E) = \frac{1}{12 a^{1/4} (E - \Delta)^{5/4}} \exp \left\{ 2 \sqrt{a(E - \Delta)} \right\}, \tag{22}$$

gives much better results (see e.g., Dilg et al., 1973) than the simple form given by eq. (7) when both a and Δ of eq. (22) are treated as free parameters. The fact we need the parameter Δ is an indication of the presence of two-body residual interaction. With a purely one-body Hamiltonian, assumed in the derivation of the Fermi-gas model, the zeropoint of the energy scale for the A-particle system is determined using eq. (18) by filling up the lowest available single-particle states. It is well known from other nuclear structure studies that the true ground state energy of a nucleus is depressed below this value because of the two-body residual interaction;

Fig. X-3.   State density
calculated for ^{56}Fe with Nils-
son and Seeger single-particle
energies taken from Haq and
Wong (1980). The combina-
torial results give the exact
state densities for a one-body
Hamiltonian. The statistical
spectroscopy results are calculated by grouping the 62
orbits into 16 shells, and the
Fermi gas results are obtained
using eq. (7) with $a = 7.2$
MeV^{-1}. The smoothed single-
particle energy curve demonstrates the effects of fluctuations in the single-particle
spectrum.

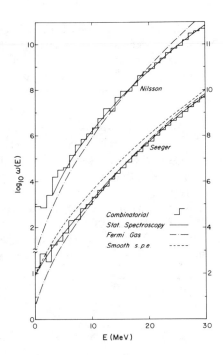

the introduction of the parameter Δ is an easy way to incorporate this
effect semi-empirically.

There is no difficulty in principle to include $H(2)$ in state density
studies by, say, configuration averaging. However, for the large spaces
required for state density studies, the calculation becomes extremely time
consuming, as in the combinatorial method. Furthermore, the fact that
the back shifted Fermi gas formula is adequate to explain a large amount
of data implies that the effect of $H(2)$ on $\omega_A(E)$ is global in nature and
does not affect the details of the energy dependence. Hence, it should
be possible to introduce a simpler scheme to account for the presence of
two-body residual interaction in state densities.

One such approach was taken by Haq and Wong (1982). The basic idea
can be described in the following way. So far in the derivation of eq. (19)
we have assumed that all the single-particle states are eigenstates of the
Hamiltonian, true for a purely one-body Hamiltonian. In the presence
of $H(2)$, at most one or two such states near the Fermi energy can be
sharp ones; all other single-particle states are superpositions of eigenstates
and therefore are distributed over some range of energy. Furthermore, the
width of the single-particle strength distribution is expected to increase
with excitation energy. Experimentally it is known that the strengths of
the highly excited single-particle states are so diffuse that it is difficult to
identify them.

As a model for state densities and related quantities, we assume that the strength of each single-particle state has a Gaussian distribution centered at the energy x_r given by $H(1)$ alone, and with a variance linearly proportional to the centroid energy, $\sigma_r^2 = \varsigma x_r$. For single-particle state r, the strength distribution is then

$$w_r(\epsilon) = \frac{1}{\sqrt{2\pi\varsigma x_r}} \exp - \frac{(\epsilon - x_r)^2}{2\varsigma x_r}, \qquad (23)$$

where the proportionality constant ς, with the dimension of energy, is related to the strength of the two-body residual interaction. The value of ς is to be determined by a separate calculation. The model is both physically reasonable and mathematically convenient to apply.

The strength distribution of a many-particle configuration m is the convolution of those of the occupied single-particle states and is therefore Gaussian in shape with centroid

$$X_m = \sum_r m_r \, x_r, \qquad (24)$$

and variance

$$\sigma_m^2 = \sum_r m_r \sigma_r^2 = \sum_r m_r \varsigma x_r = \varsigma X_m. \qquad (25)$$

Eqs. (24) and (25) are exact for Gaussian single-particle strength distributions; otherwise the central limit theorem can be invoked to obtain the same result.

When all configuration strength distributions have a finite variance, the state density at a given energy E becomes an integral over all the contributing configurations,

$$w_A(E) = \sum_m \frac{d_m(X)}{\sqrt{2\pi\varsigma X_m}} \exp - \frac{(E - X_m)^2}{2\varsigma X_m}$$

$$\longrightarrow \int_0^\infty \frac{dx}{\sqrt{2\pi\varsigma X}} \exp - \frac{(E - X)^2}{2\varsigma X} \, dX, \qquad (26)$$

where d_m is the number of states in configuration m with centroid at X_m and variance ςX_m. In the limit when the summation over configurations can be replaced by an integration, d_X, the number of states with centroid at X, can be approximated by using a Fermi gas model, or by a Darwin-Fowler method (Haq and Wong, 1982) if a set of realistic single-particle energies is to be used. In the absence of $H(2)$, the distribution of each configuration reduces to a delta function and we recover eq. (20).

It should be noted that because of the finite width of the distributions of all configurations, the strength is no longer zero below the lowest configuration centroid at $X = 0$. For a one-body Hamiltonian, the ground state of the system is at $X = 0$ which is taken as the zeropoint of the energy scale for the many-particle system. In the presence of $H(2)$, the ground state is pushed below $X = 0$ because there is strength below it. The new ground state position may be determined, for example, by the method of the previous section, or by empirically fitting the state density at high excitation energy, as done by Haq and Wong (1982).

The ratio ς between the variance and the centroid of a configuration depends on the nature of the residual interaction. It was argued by French and Kota (1983) that only certain unitary rank parts of the residual interaction can contribute to the spreading of the configuration distribution, and they have given an example where an increase in the variance with increasing centroid energies is obtained for a realistic interaction. This provides a direct link between the state density and the nuclear Hamiltonian that was lacking for some time.

The partition function, $Z(\alpha, \beta)$, of eq. (10) is also of physical interest by itself. If we measure the temperature T in energy units (i.e., include the Boltzmann constant k in the definition of T) and ignore the chemical potential, then $Z(\alpha, \beta) \to Z(T)$. The internal energy,

$$U(T) = -\frac{\partial}{\partial \beta} \ln Z(\alpha, \beta) = T^2 \frac{\partial}{\partial T} \ln Z(T), \qquad (27)$$

determines the average amount of energy that can be stored in the excited states of a nucleus at temperature T. The energy $U(T)$ is an important quantity in astrophysics since it determines the amount of mechanical energy removed from the collapse of a massive star prior to a supernova explosion.

Physically we expect that at extremely high temperatures, the nucleus will break up and the nuclear partition function will vanish. However the expression for $\omega(E)$ given by eq. (7), and consequently the partition function derived from it, diverges as $T \to \infty$ (Bethe et al., 1979). This property is due to the infinite dimensional space assumed for the nucleus in deriving eq. (7). However, a nucleus cannot survive as a bound system at infinite excitation energy and the state density of a nucleus must decrease when the temperature is sufficiently high. The proper state density should therefore be calculated in a finite space with a finite total number of bound states and bound states imbedded in the continuum, and their contributions to $Z(T)$ approach some constant value asymptotically.

Since statistical spectroscopy operates in a finite space, it provides a natural way to obtain the state density and nuclear partition function at high temperatures (Wong, Zhao, and Zhu, 1980). Experimental guidance is required to determine the proper size of the space to be used and perhaps astrophysical studies may be of help here.

X.3 Level density and spin cut-off factor

In general, the spin J and isospin T are assumed to be good quantum numbers of a nuclear state and all states differing only in M and Z, the projections of J and T on the quantization axes, are degenerate in energy. Instead of the density of states $\omega_A(E)$, we can consider the density of levels $\rho(E)$. For a given (J, T), the two quantities are related by the degeneracy of the level,

$$\omega_{JT}(E) = (2J + 1)(2T + 1)\, \rho_{JT}(E). \tag{28}$$

More generally,

$$\omega_A(E) = \sum_{JT} \omega_{JT}(E) = \sum_{JT} (2J + 1)(2T + 1)\, \rho_{JT}(E), \tag{29}$$

and

$$\rho_A(E) = \sum_{JT} \rho_{JT}(E). \tag{30}$$

The ratio between the two quantities is given by the spin cut-off factor $\sigma_J^2(E)$ in the form

$$\rho_A(E) = \frac{\omega_A(E)}{\sqrt{2\pi\sigma_J^2(E)}}. \tag{31}$$

The factor $\sqrt{2\pi\sigma_J^2(E)}$ is therefore the average number of states per level at excitation energy E.

We can relate the spin cut-off factor $\sigma_J^2(E)$ to the variance of the state density distribution as a function of M at energy E. In a dilute system, $N \gg m \gg 1$, the effect of the Pauli principle is unimportant, as we have noted earlier. The distribution of states as a function of M is therefore Gaussian in virtue of the central limit theorem. Furthermore, since the distribution is symmetric around $M = 0$, it is zero-centered. Ignoring the isospin for the time being to simplify the argument, the distribution at a given energy can be written as

$$p(M) = \frac{1}{\sqrt{2\pi\langle M^2\rangle}} \exp -\frac{M^2}{2\langle M^2\rangle}. \tag{32}$$

Analogous to eq. (VIII-3), the distribution of states as a function of J can be found from the difference between $p(M = J)$ and $p(M = J + 1)$,

$$
\begin{aligned}
q(J) &= p(M = J) - p(M = J + 1) \\
&\approx -\left(\frac{\partial p(M)}{\partial M}\right)_{M=J+1/2} \\
&= \frac{1}{\sqrt{2\pi\langle M^2\rangle}} \frac{2J + 1}{2\langle M^2\rangle} \exp -\frac{(J + 1/2)^2}{2\langle M^2\rangle}.
\end{aligned}
\tag{33}
$$

Since the degeneracy of a level with spin J is $(2J+1)$, the average number of states per level at energy E is

$$\langle (2J+1) \rangle \approx \frac{\int (2J+1)\, q(J)\, dJ}{\int q(J)\, dJ} = \sqrt{2\pi \langle M^2 \rangle}, \qquad (34)$$

where we have taken the integral over J between 0 and ∞, valid in large spaces. Comparing eq. (34) with eqs. (29) and (31), we can make the identification

$$\sigma_J^2(E) = \langle M^2 \rangle.$$

The spin cut-off factor is therefore the variance of the distribution of states as a function of M.

Since $\sigma_J^2(E)$ is related to the average value of J at energy E, it is needed in a variety of nuclear reaction calculations involving compound nuclei. In a single j-orbit,

$$\sigma_J^2 = \frac{1}{N} \sum_{M=-J}^{J} M^2 = \frac{1}{3} j(j+1),$$

where $N = 2j + 1$. For a space consisting of several orbits,

$$\sigma_J^2(m) = \frac{1}{3N} \sum_r j_r (j_r + 1)(2j_r + 1), \qquad (35)$$

where $N = \sum_r N_r = \sum_r (2j_r + 1)$. For a given configuration, σ_J^2 is a constant independent of the energy since each orbit has a definite number of particles. The energy variation of $\sigma_J^2(E)$ in the complete space comes therefore from the difference in the relative contributions from various configurations at each energy. From this argument we see that $\sigma_J^2(E)$ can only vary slowly with energy, as is observed experimentally (Grimes et al., 1978). By the same token, we can obtain the value of the spin cut-off factor easily from configuration averaging. Since the relative intensity of configuration m is given by $I_m(E)$, we obtain

$$\sigma_J^2(E) = \frac{1}{I_m(E)} \sum_m I_m(E) \sigma_J^2(m), \qquad (36)$$

where $I_m(E) = \sum_m I_m(E)$ and $\sigma_J^2(m)$ is given by eq. (35).

We can also obtain $\sigma_J^2(E)$ by scalar averaging using the orthogonal polynomial expansion of strength distributions given in Chapter III. The interest here is in the variations of M^2 as a function of energy. Since the operator associated with M^2 is J_z^2, the spin cut-off factor can be expanded in the form given by eq. (III-31),

$$\sigma_J^2(E) = \sum_\mu \langle J_z^2\, P_\mu(H) \rangle\, P_\mu(E). \qquad (37)$$

If a one-body Hamiltonian is used, the necessary input moments up to order 8, adequate to obtain $P_\mu(E)$ up to $\mu = 4$ with eq. (III-29), are given by

$$
\begin{aligned}
\langle J_z^2 \rangle =\ & P(2,1)A_0 \,, \\
\langle J_z^2 H(1) \rangle =\ & [P(3,1) - P(3,2)]A_1 \,, \\
\langle J_z^2 H(1)^2 \rangle =\ & [P(4,1) - 4P(4,2) + P(4,3)]A_2 + P(4,2)A_0 A_2 \,, \\
\langle J_z^2 H(1)^3 \rangle =\ & [P(5,1) - 11P(5,2) + 11P(5,3) - P(5,4)]A_3 \\
& + [P(5,2) - P(5,3)]3A_1 A_2 + [P(5,2) - P(5,3)]A_0 A_2 \,, \\
\langle J_z^2 H(1)^4 \rangle =\ & [P(6,1) - 26P(6,2) + 66P(6,3) - 26P(6,4) + P(6,5)]A_4 \\
& + [6P(6,2) - 24P(6,3) + 6P(6,4)]A_2 A_2 \\
& + [4P(6,2) - 8P(6,3) + 4P(6,4)]A_1 A_3 + 3P(6,3)A_0 A_2^2 \\
& + [P(6,2) - 4P(6,3) + P(6,4)]A_0 A_4 \,, \qquad (38)
\end{aligned}
$$

where

$$
A_q = \frac{1}{3} \sum_r j_r(j_r + 1)(2j_r + 1)(\epsilon_r - \bar{\epsilon})^q \,,
$$

$$
\Lambda_q = \sum_r N_r(\epsilon_r - \bar{\epsilon})^q \,, \quad \text{and} \quad P(p,t) = \binom{N}{m}^{-1}\binom{N-p}{m-t} \,.
$$

Once we have the expressions for the polynomials $P_\mu(E)$, the expansion coefficients $\langle J_z^2 P_\mu(H) \rangle$ in eq. (37) can be written in terms of $\langle J_z^2 H^\nu \rangle$ using the expressions provided by eq. (38). Hence eqs. (21) and (38) alone are adequate to calculate the spin cut-off factor for a one-body Hamiltonian up to fourth order in scalar averaging.

Scalar averaging of this kind is not expected to be accurate enough in the large spaces needed for level densities. On the other hand, a configuration averaging based on shells using a one-body Hamiltonian can be carried out with the information provided here in a way similar to the calculation carried out for state densities in the previous section. It is likely that such an approach may be adequate for the spin cut-off factor. In a test study, the calculated results for level densities and spin cut-off factors (Haq and Wong, 1980) agreed with experiment except for a shift in the energy scale of about 3 MeV. This was attributed to two-body residual effects which may be accounted for in a similar way as for state densities (Haq and Wong, 1982).

There are two sources of experimental data of level densities available for comparison. In the low energy region, where the individual levels are known together with their (J, T) assignments, an actual counting of the states can be carried out to obtain both $\rho_A(E)$ and $\sigma_J^2(E)$. Since the level spacings here are large, it is not the region of primary interest for level

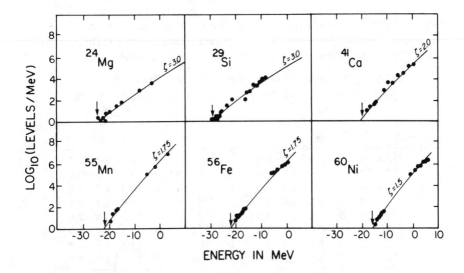

Fig. X-4. Level densities for a few typical light nuclei taken from Haq and Wong (1982). $H(1)$ is included in the study through a saddle point integration and $H(2)$ by the approximate model described in section 3.

density studies; however, the data do provide a useful bench mark test of the calculated results as can be seen in Fig. X-4. At higher excitation energies, the individual levels are no longer resolved; the level density must be deduced from certain reaction cross-section data in a weakly model dependent way (Huizenga and Moretto, 1972). The results provide a second check of the calculated results.

X.4   Occupancy

Many physical quantities are to a first approximation sensitive only to the number of particles in each orbit. Just as in the case of the spin cut-off factor, the occupancy of each orbit is constant within a given configuration, and the variations of $m_r(E)$, the number of particles in orbit r as a function of energy E, come from the differences in the contributions from different configurations. In the limit of Gaussian configuration strength distributions, we obtain the result

$$m_r(E) = \frac{1}{I_m(E)} \sum_m \langle \hat{n}_r \rangle^{(m)} d_m \rho_m(E)$$

$$\xrightarrow[\text{Gaussian}]{} \frac{1}{I_m(E)} \sum_m \langle \hat{n}_r \rangle^{(m)} \frac{d_m}{\sigma_m \sqrt{2\pi}} \exp -\frac{(E - C_m)^2}{2\sigma_m^2} , \quad (39)$$

where $\langle \hat{n}_r \rangle^{(m)}$ is the occupancy of orbit r in configuration m and $I_m(E) = d_m \rho(E)$. The calculated results for the ground state occupancies are given in Table X-2 for some of the ds-shell nuclei listed in Table X-1.

Table X-2 Ground state occupancies of some ds-shell nuclei
(CW particle interaction for $A \leq 28$ and hole interaction for $A > 28$)

A	2J	2T	$d_{5/2}$	$d_{3/2}$	$s_{1/2}$	A	2J	2T	$d_{5/2}$	$d_{3/2}$	$s_{1/2}$
^{20}Ne	0	0	2.9	0.7	0.4	^{29}Si	1	1	8.6	2.6	1.8
^{21}Ne	3	1	3.8	0.6	0.6	^{30}P	2	0	9.2	2.8	2.0
^{22}Na	6	0	4.6	0.8	0.6	^{31}P	1	1	10.0	2.8	2.2
^{23}Na	3	1	5.3	0.9	0.8	^{32}S	0	0	10.8	2.8	2.4
^{24}Mg	0	0	5.1	1.9	1.0	^{33}S	3	1	11.0	3.2	2.8
^{25}Mg	5	1	6.7	1.3	1.0	^{34}Cl	6	0	11.3	3.7	3.0
^{26}Al	10	0	7.5	1.4	1.1	^{35}Cl	3	1	11.6	4.1	3.3
^{28}Si	0	0	8.3	2.2	1.5	^{36}Ar	0	0	11.8	4.6	3.6

Experimentally the ground state occupancy is obtained via the non-energy weighted sum rule quantities for one-nucleon transfer reactions. The spectroscopic factor S_r for *picking up* a particle in orbit r from a target state $|m\Gamma\alpha\rangle$ to a final state $|(m-1)\Gamma'\alpha'\rangle$ is given by

$$S_r^{(-)} = \frac{1}{[\Gamma]} |\langle (m-1\Gamma'\alpha' \| B^r \| m\Gamma\alpha \rangle|^2.$$

On summing over all possible final states in the $(m-1)$-nucleon space, we obtain the non-energy weighted sum rule quantity for a pickup reaction,

$$G_0^{(-)}(r) = \frac{1}{[\Gamma]} \sum_{\Gamma'\alpha'} |\langle (m-1)\Gamma'\alpha' \| B^r \| m\Gamma\alpha \rangle|^2$$

$$= \frac{1}{[\Gamma]} \sum_{\Gamma'\alpha'} (-1)^{\Gamma'+r+\Gamma} \langle m\Gamma\alpha \| A^r \| (m-1)\Gamma'\alpha' \rangle$$

$$\times \langle (m-1)\Gamma'\alpha' \| B^r \| m\Gamma\alpha \rangle$$

$$= \left[\frac{r}{\Gamma}\right]^{\frac{1}{2}} \sum_{\Gamma'\alpha'} \begin{Bmatrix} \Gamma & r & \Gamma' \\ r & \Gamma & 0 \end{Bmatrix} \langle m\Gamma\alpha \| A^r \| (m-1)\Gamma'\alpha' \rangle$$

$$\times \langle (m-1)\Gamma'\alpha' \| B^r \| m\Gamma\alpha \rangle$$

$$= \left[\frac{r}{\Gamma}\right]^{\frac{1}{2}} \langle m\Gamma\alpha \| (A^r \times B^r)^0 \| m\Gamma\alpha \rangle = m_r, \qquad (40)$$

where we have used eq. (C-16) to change the reduced matrix elements of B^r to those of A^r and eq. (C-20) to recouple the reduced matrix elements.

$G_0^{(-)}(r)$ therefore gives the number of particles in orbit r of the target state since $[r]^{\frac{1}{2}}(A^r \times B^r)^0$ is the number operator.

The non-energy weighted sum rule quantity for stripping reactions, on the other hand, accounts for the number of holes in an orbit,

$$
\begin{aligned}
G_0^{(+)}(r) &= \frac{1}{[\Gamma]} \sum_{\alpha'\Gamma'} S_r^{(+)} \\
&= \frac{1}{[\Gamma]} \sum_{\alpha'\Gamma'} |\langle (m+1)\Gamma'\alpha' \| A^r \| m\Gamma\alpha \rangle|^2 \\
&= \frac{1}{[\Gamma]} \sum_{\alpha'\Gamma'} |\langle (N-m-1)\Gamma'\alpha' \| B^r \| (N-m)\Gamma\alpha \rangle|^2 \\
&= \left[\frac{r}{\Gamma}\right]^{\frac{1}{2}} \langle (N-m)\Gamma\alpha \| (A^r \times B^r)^0 \| m\Gamma\alpha \rangle = (N_r - m_r), \quad (41)
\end{aligned}
$$

where a particle-hole transformation is used to obtain the second step. Thereafter, the derivation is the same as that in obtaining eq. (40).

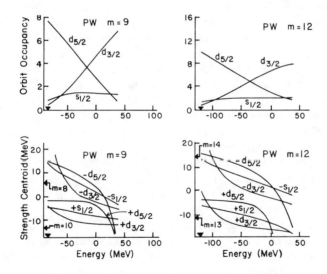

Fig. X-5.   Occupancy and centroids for stripping and pickup strengths calculated with a Preedom-Wildenthal (1972) PW interaction (Draayer *et al.*, 1975). Except at both ends of the scale, the energy dependences are fairly linear.

For the excited states, an experimental determination of the occupancy by the same sum rule methods as used for the ground state is not possible. The quantity is nevertheless of interest. Two examples, taken from the

results of Draayer *et al.* (1975), are shown in Fig. X-5. We see that the
energy variation of $m_r(E)$ is smooth, as expected from the fact that many
configurations are contributing at a given energy. The difference in the
energy dependence between orbits is primarily a consequence of the single-
particle energies involved. In the ds-shell, the $d_{5/2}$ single-particle orbit is
the lowest one and consequently its occupancy is the largest in the ground
state region. The $d_{3/2}$ orbit, on the other hand, has a small occupancy
at low excitations since its single-particle energy is the highest one. The
slopes of the energy dependences are essentially determined by the fact
that the relative occupancies between $d_{5/2}$ and $d_{3/2}$ orbits are reversed at
the high energy end.

It is perhaps more instructive to examine the problem from the point
of view of scalar averaging. Again, using the polynomial expansion of
eq. (III-31), we obtain

$$m_r(E) = \sum_\mu \langle \hat{n}_r P_\mu(H) \rangle P_\mu(E)$$

$$= \left\{ m \frac{N_r}{N} + \langle \hat{n}_r \cdot (H - C)/\sigma \rangle \frac{E - C}{\sigma} + \cdots \right\}. \qquad (42)$$

The first term is the number of particles in orbit r averaged over the entire
m-particle space. The second term, which gives the linear energy dependence, is given by the correlation between $\hat{n}_r$ and H. Since $\hat{n}_r$ is a unitary
rank $\nu = 1$ operator (in scalar averaging), only the $\nu = 1$ part of the
Hamiltonian can contribute to $\langle \hat{n}_r \cdot (H - C)/\sigma \rangle$. For a one-body Hamiltonian, the $\nu = 1$ part depends on $\tilde{\epsilon}_r$, the traceless single-particle energy of
eq. (V-23): it is negative for $d_{5/2}$ and positive for $d_{3/2}$; these factors are
reflected in the slopes of the two occupancies. Since the $s_{1/2}$ orbit is more
or less at the average value of the three single-particle energies, $\tilde{\epsilon}_{s_{1/2}}$ is
essentially zero and this explains the lack of energy dependence of the $s_{1/2}$
orbit. The two-body interaction enters into the linear-energy dependence
only through the induced single-particle energy term, given by eq. (V-32).
It modifies the values of the slopes from those given by $H(1)$ alone but the
general trend remains the same.

From the central limit theorem, we expect the energy dependence of
$m_r(E)$ to be linear. The results of shell model calculations are in agreement
with this expectation except near the two ends of the spectrum. This is also
seen in the configuration-averaging results shown in Fig. X-5 and implies
that the higher-order terms in eq. (42) are not very important except near
the ground state region (Draayer, French, and Wong, 1977).

The linear energy-weighted sum rule quantity for pickup reactions,

$$G_1^{(-)}(r, E) = \langle E \| (A^r \times H \times B^r)^0 \| E \rangle, \qquad (43)$$

gives the centroid of pickup strengths, $G_1^{(-)}(r, E)/G_0^{(-)}(r, E)$, from a *target*
state at energy E. Except near the two ends of the spectrum, the energy
dependence is again observed to be linear as shown in Fig. X-5.

The quadratic energy-weighted sum rule,

$$G_2^{(-)}(r, E) = \langle E \| \left(A^r \times H^2 \times B^r \right)^0 \| E \rangle, \tag{44}$$

is related to the variance of the one-nucleon transfer strength distribution,

$$V^{(-)}(r, E) = \frac{\langle E \| \left(A^r \times H^2 \times B^r \right)^0 \| E \rangle}{\langle E \| \left(A^r \times B^r \right)^0 \| E \rangle} - \left\{ \frac{\langle E \| \left(A^r \times H \times B^r \right)^0 \| E \rangle}{\langle E \| \left(A^r \times B^r \right)^0 \| E \rangle} \right\}^2$$

$$= \frac{G_2^{(-)}(r, E)}{G_0^{(-)}(r, E)} - \left\{ \frac{G_1^{(-)}(r, E)}{G_0^{(-)}(r, E)} \right\}^2. \tag{45}$$

The value of $V^{(-)}(r, E)$ turns out to be small and constant both in shell model and statistical spectroscopy results. Physically this means that the one-nucleon transfer strength from a target state at energy E tends to go to a narrow region in the daughter nucleus at E'. As we move up in energy E, the value of E' also increases. If the strength distribution, $R(E', E)$, is displayed as a function of both E and E', the plot will show the strength concentrated in a narrow ridge along $E \approx E'$, seen also in the electric quadrupole transition strength distribution to be described in the next section. The existence of a narrow ridge is not surprising in one-nucleon transfer reactions since the operator removes only a particle from, or adds a particle to, the initial state without affecting other nucleons. A group of states in a given energy region are likely to have similar structure, at least as far as their orbital occupancies are concerned. As a result, there is a tendency for reactions involving a single nucleon to connect two narrow regions in energy and thus lead to a narrow ridge in the strength distribution.

X.5   E2 and other electromagnetic transition strengths

The electric quadrupole (E2) operator is more strongly correlated with the nuclear Hamiltonian than the number operator. As a result, we expect a more complicated strength distribution including strongly collective transitions, which are however of no interest to us here. To compare the calculated E2 strength distribution with experimental data, we need measurements of the transition strengths from all the states in a given energy region to all the states in another energy region. Two types of data are available. At low energies, where the states are well isolated, all the individual transitions can be measured. These transitions are interesting for a variety of purposes, but they do not cover a sufficiently large region (in spacing units) to be useful in statistical spectroscopy. A second type of transition are the giant resonances. Since these are transitions with the final state restricted to the ground state, they do not provide a complete picture of the strength distribution $R(E', E)$; we shall however return to them later.

Since there is no suitable experimental data, we shall make comparisons of distribution results with a set of *simulated data* from a shell model calculation of the E2 transitions between the subspaces $(J^\pi, T) = (0^+, 0)$ and $(2^+, 0)$ in the ds-shell with six active particles (Draayer, French, and Wong, 1977). The shell model values are shown in Fig. X-6 in the form of a three-dimensional display. The initial and final states are labelled by the excitation energies E and E' measured from the lowest states of each J value. The transition between a pair of states is represented by a spike with the height proportional to the strength. The Hamiltonian used consists of a one-body part given by the ^{17}O single-particle energies and a renormalized two-body interaction (Kuo, 1967).

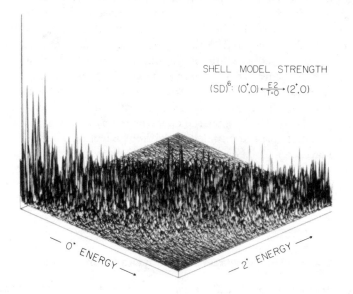

Fig. X-6. E2 transition strengths between 71 $(J^\pi, T) = (0^+, 0)$ and 307 $(2^+, 0)$ states of $(ds)^6$. The strength of each of the 21,797 transitions is represented by the height of the spike with the strongest one being 39 $e^2 fm^4$ (between 0_1^+ and 2_5^+).

As usual, the state-to-state variations of the transition strength are not of interest here and we shall smooth out the fluctuations from the beginning by a local average. Each point of the plot in Fig. X-7 is the average value of all the strengths in a square grid area with either 4 or 8 MeV on the side as indicated. The choice of the grid size is not arbitrary: it must be large enough so that the state-to-state fluctuations can be removed and yet small enough to retain the interesting secular behaviour. The plot shows clearly that the strength is concentrated along the $E \approx E'$ ridge with a fairly constant width. The total transition strength starting from a given state is, however, much larger on the average at low energies, a result well known from the abundance of strong low-lying E2 strengths in many nuclei, in particular the even-even ones in the ds-shell.

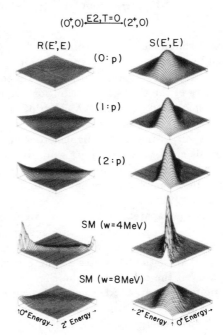

$(0^+, 0) \xrightarrow{\text{E2,T=0}} (2^+, 0)$

Fig. X-7. E2 strength between $(J^\pi, T) = (0^+, 0)$ and $(2^+, 0)$ for $(ds)^6$. Two smoothed shell-model values are shown, one with a square grid of 4 MeV on a side and the other with 8 MeV. The configuration results are identified by the maximum polynomial order used. With terms up to second order, $(2:p)$, the results are comparable to the exact values smoothed with a grid between 4 to 8 MeV on a side.

We do not expect that an electromagnetic transition strength distribution, such as the E2 example here, can be accounted for by a low-order scalar polynomial theory. On the other hand, configuration averaging should be able to provide an adequate description of the smoothed shell model results with low-order polynomials. In Fig. X-7, three sets of distribution results are shown for the different maximum polynomial orders used in eq. (III-41) to obtain $R(E', E)$. With only $\mu = \nu = 0$ (0:p in Fig. X-7), we obtain

$$R^{00}(E', E) = \frac{1}{I(E)I(E')} \sum_{mm'} \frac{1}{d_{m'}} I_m(E) \, I_{m'}(E') \, \langle \hat{O}^\dagger \hat{O} \rangle^{m(m')}.$$

In contrast to a zeroth-order scalar theory, we have here some variation of $R(E', E)$ as a function of the energies because different pairs of configurations m and m' have different average strengths. On the other hand, without some appropriate energy variations within each $R(m'E', mE)$, there is a marked lack of structure in the energy dependence of $R^{00}(E', E)$ compared with the smoothed shell model results.

On including all the terms up to $\mu = \nu = 1$ (1:p in Fig. X-7),

$$R^{11}(E', E) = R^{00}(E', E) + \frac{1}{I(E)I(E')} \sum_{mm'} \frac{1}{d_{m'}} I_m(E) \, I_{m'}(E')$$

$$\times \left\{ \langle \hat{O}^\dagger H \hat{O} \rangle^{m(m')} + \langle \hat{O}^\dagger \hat{O} H \rangle^{m(m')} + \langle \hat{O}^\dagger H \hat{O} H \rangle^{m(m')} \right\}, \quad (46)$$

a more pronounced energy concentration along the ridge than $R^{00}(E', E)$ is seen but is still inferior compared with the shell model results averaged even by an 8 MeV grid. With terms up to $\mu = \nu = 2$ (2:p in Fig. X-7), structures of the strength distribution begin to emerge that go beyond the results smoothed with the 8 MeV grid. However, to reproduce the results for a 4 MeV grid, polynomials higher than the second order are required. It is not surprising that a higher-order polynomial theory is needed to fit the result smoothed by a finer grid. In fact, a formal connection can be made between the grid size for averaging the shell model results and the maximum order of polynomial required to fit it. We shall, however, not pursue this subject here.

It is perhaps easier to see the significance of certain aspects of the strength distribution through sum rule quantities. If we restrict the transitions to be between 0^+ and 2^+ states, we have only one set of sum rule quantities with the 0^+ states as the initial space. Starting from a given 2^+ state, on the other hand, the E2 operator leads to final states with J^π ranging from 0^+ to 4^+: consequently, sums with the final states restricted to 0^+ states are not proper sum rule quantities.

The non-energy weighted sum rule quantity $G_0(E)$, shown in Fig. X-8, has a high concentration of strength at the low energies, decreasing rapidly with increasing excitation energy. The general trend is well described by a linear energy dependence related to the correlation coefficient between the Hamiltonian H and the square of the E2 operator, usually expressed in the form of a $\hat{Q} \cdot \hat{Q}$ operator. The large (negative) slope is expected from the strong correlation of $\hat{Q} \cdot \hat{Q}$ with the ds-shell Hamiltonian. In fact, the correlation is strong enough that some quadratic energy dependences, produced by the higher-order correlations of $\hat{Q} \cdot \hat{Q}$ with H, are easily noticeable in the plot. For the renormalized interaction used in the example, the correlation coefficient, described in the next section, is -0.55 in the $(ds)^6$ space. Roughly the same value is also obtained for all the other realistic Hamiltonians in the ds-shell.

The position of the ridge in the $R(E', E)$ plot is related to the ratio $G_1(E)/G_0(E)$, the linearly energy-weighted sum divided by the non-energy weighted one. The energy dependence of $G_1(E)/G_0(E)$ is essentially a linear one reflecting the smooth energy variation of the ridge in the $R(E', E)$

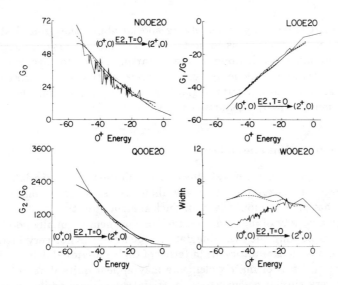

Fig. X-8. Sum rule quantities for E2 transitions from $(0^+, 0)$ to $(2^+, 0)$. Solid lines are the second-order configuration results and dashed lines are the fourth-order scalar results. The strength width is small, around 5 MeV, and constant.

distribution. The variance of the strength distribution, $G_2(E)/G_0(E) - \{G_1(E)/G_0(E)\}^2$, involves the quadratic sum rule quantity $G_2(E)$ and its values are small here, as can be expected from the narrowness of the $R(E', E)$ ridge. However, it may be difficult to expect good accuracy in calculating the variance using a truncated polynomial expansion for each of the three sum rule quantities $G_0(E)$, $G_1(E)$, and $G_2(E)$ involved in the expression because of the cumulative error arising from taking the difference between the ratios of two approximate results.

Giant resonances are collective electromagnetic transitions from the ground state to a region where the individual excited states are usually not resolved. Besides the well known giant dipole resonances, giant quadrupole and other multipole resonances have been identified experimentally (see *e.g.*, Betrand, 1981). Since the space for the final states may involve several major shells, it is unlikely that microscopic calculations can be carried out in such large spaces.

The experimentally observed resonances are those built upon the target ground state. Leaving aside the question of actual observation, we can consider the same type of resonances built upon the excited states of the target. (In fact, at high stellar temperatures there are enough excited nuclei in equilibrium that resonances built upon the excited states must be included.) Unless there is something special about the ground state for electromagnetic transitions to states at very high excitation energies, we expect resonances built upon different starting states to vary only slowly with the starting state energy. In such cases, statistical spectroscopy can be used as a tool to understand giant resonances by studying the strength distribution for low-lying starting states.

X.6   Correlation coefficient

It often happens that several effective Hamiltonians are available in the same space. For example, there are about ten such Hamiltonians in the ds-shell that are commonly used for nuclear structure studies. They are obtained by a variety of methods and each one of them is equally good on the whole in explaining the available data. In general it is not very instructive to compare these Hamiltonians in terms of their individual defining matrix elements: two sets of matrix elements may appear quite different and yet produce very similar eigenvectors. A trivial example is adding a constant to all the single-particle energies: the only effect is to shift all the eigenvalues by m times the constant. At the same time, it is not easy to draw any meaningful conclusion by comparing the eigenvectors obtained with different Hamiltonians since the results calculated with these eigenvectors tend to fit different sets of data with different degrees of success.

A good way to assess the differences and similarities between two Hamiltonians is to have a *global* measure. Let the number of defining matrix elements be d. Each Hamiltonian can then be represented by a vector in a d-dimensional space and the projection of the vector on a given axis is the value of the corresponding defining matrix element. Different Hamiltonians appear as different vectors in this space.

The correlation coefficient Ξ_{ab} between two vectors a and b in such a multi-dimensional space is defined as the cosine of the angle between them,

$$\Xi_{ab} \equiv \frac{a \cdot b}{|a|\,|b|}\,, \tag{47}$$

where $|\Xi_{ab}| \leq 1$. It is equal to $+1$ if a and b are parallel, -1 if they are parallel but pointing in opposite directions, and 0 if they are perpendicular to each other.

To compare two Hamiltonians H and K by means of their correlation coefficient, both *vectors* must first be brought to the same origin. This can be achieved by removing the centroids, C_H and C_K respectively, from each Hamiltonian and making them traceless. The correlation coefficient between H and K is then given by

$$\Xi_{HK} = \frac{\langle (H - C_H) \cdot (K - C_K) \rangle}{\sigma_H \cdot \sigma_K}\,, \tag{48}$$

where $\sigma_H^2 = \langle (H - C_H)^2 \rangle$ is the norm of the vector H, and σ_K^2 is the corresponding quantity for K.

Since the meaning of an average depends on the group of states over which the trace is taken, we can define different correlation coefficients by using different averaging techniques, scalar, configuration, fixed-JT, etc. Here we shall be mainly concerned with the scalar correlation coefficient, i.e., taking the average of eq. (48) over all the m-particle states.

In the ds-shell, a comparison of the different available effective Hamiltonians has been made by Potbhare (1977). It is found that all the reasonable ones have correlation coefficients between each other greater than 0.9. This means that, in the (63+3)-dimensional space that describes the one- plus two-body Hamiltonian in the ds-shell, the *vectors* representing all the effective Hamiltonians are clustered in a narrow *cone* with half angle less than 13°. Indeed, all these Hamiltonians tend to produce very similar over-all results even though they may differ on a particular quantity. Furthermore, the fact that ds-Hamiltonians obtained from different sources are strongly correlated implies that the "true" ds-effective Hamiltonian is likely to be found within this *cone*.

Table X-3 Correlation coefficients between some ds-shell Hamiltonians

m	CW particle	Kuo 1967	Rosenfeld	$\hat{Q} \cdot \hat{Q}$	interaction
6	10.49*	0.99	0.96	-0.50	CW particle
		9.83*	0.95	-0.55	Kuo 1967
			8.96*	-0.46	Rosenfeld
				66.51*	$\hat{Q} \cdot \hat{Q}$
9	11.77*	0.98	0.95	-0.51	CW particle
		11.16*	0.95	-0.56	Kuo 1967
			10.05*	-0.47	Rosenfeld
				84.64*	$\hat{Q} \cdot \hat{Q}$
12	12.96*	0.98	0.95	-0.50	CW particle
		12.05*	0.94	-0.55	Kuo 1967
			10.60*	-0.47	Rosenfeld
				90.98*	$\hat{Q} \cdot \hat{Q}$

*Value of σ_H.

Table X-3 lists the scalar correlation coefficients between four typical Hamiltonians in the ds-shell. As mentioned in section 1, the CW particle interaction (Wildenthal and Chung, 1979) is obtained by fitting the 63+3 defining matrix elements to certain ds-shell data. The Kuo (1967) interaction, on the other hand, starts with a realistic nucleon-nucleon interaction and accounts for the effects of truncating the Hilbert space by renormalizing the two-body matrix elements. Historically, before large computers were employed for the numerical calculations, the Rosenfeld interaction (see *e.g.*, Eisenberg and Griener, 1972, p. 248) was often used because of its simplicity. The $\hat{Q} \cdot \hat{Q}$ force, given by eq. (IX-25) in terms of the bilinear $SU(3)$ Casimir operator, represents a schematic interaction that has some validity in the ds-shell. From the table we see that the correlation coefficients between the effective Hamiltonians (CW particle, Kuo, and Rosenfeld) are quite large in spite of their very different origins: the correlations with the $\hat{Q} \cdot \hat{Q}$ force are smaller by comparison but the values clearly indicate that a large part of the ds-interaction is quadrupole in nature. This is well known, and it is the primary reason for the success of the $SU(3)$ scheme (Elliot, 1963) in the space.

The calculation of a correlation coefficient (eq. 48) is quite simple. It can be carried out for arbitrarily large spaces provided that the defining matrix elements of the operators are available. We can therefore apply it not only to study different effective Hamiltonians but also, for example, different sets of G-matrix elements derived directly from free nucleon-nucleon scattering. The correlation coefficients in different isospin spaces provide the relation between two different Hamiltonians as a function of T and, if the fixed-JT averaging method is used, as a function of J as well. The possibilities are numerous and have not been fully explored.

X.7 Symmetry preserving part of the Hamiltonian

The concept of correlation coefficient can also be used to separate out the symmetry preserving part of a Hamiltonian. If a group having a Casimir operator G describes an exact symmetry, then G and H commute. However, there are only few such exact symmetries. In general, we are more interested in the broken symmetries, for which G and H do not commute and the Hamiltonian contains a part that cannot be constructed out of the generators of the group. Such a Hamiltonian will in general have non-vanishing matrix elements between different irreducible representations of the groups.

It is of interest to find out how good a particular symmetry is with respect to a given Hamiltonian. The problem may be studied by resolving H into two parts, one parallel to G, which preserves the symmetry, and the remainder orthogonal to G which breaks the symmetry,

$$H = aG + X. \tag{49}$$

The symmetry breaking part, X, is defined by

$$\langle (X - C_X) \cdot (G - C_G) \rangle = 0, \tag{50}$$

where C_X and C_G are respectively the averages (or centroids) of X and G. Since

$$\langle (H - C_H) \cdot (G - C_G) \rangle = \langle \{a(G - C_G) + (X - C_X)\} \cdot (G - C_G) \rangle$$

$$= a\sigma_G^2, \tag{51}$$

we have

$$a = \frac{\sigma_H}{\sigma_G} \Xi_{HG}. \tag{52}$$

Hence a large correlation coefficient between H and G implies that the Hamiltonian contains a large symmetry preserving part, and the symmetry is then approximately a good one.

For example, we find from Table X-3 that for $G = \hat{Q} \cdot \hat{Q}$, $a = -0.50$, -0.55, and -0.46 for respectively the CW particle, Kuo, and Rosenfeld interactions at $m = 6$. Since a large part of the bilinear Casimir operator, given in eq. (IX-25), is made of $\hat{Q} \cdot \hat{Q}$, the $SU(3)$ symmetry is slightly better preserved with the Kuo interaction rather than with a Rosenfeld interaction.

X.8   Trace equivalent Hamiltonian

The separation of a Hamiltonian into symmetry-preserving and symmetry-breaking parts can be extended to chains of groups (Countee *et al.*, 1981; Halemane, Kar, and Draayer, 1978). Let $\{\hat{O}_1, \hat{O}_2, \ldots, \hat{O}_k\}$ be a set of linearly independent, mutually orthogonal operators. We can write

$$H = c_1 \hat{O}_1 + c_2 \hat{O}_2 + \cdots + c_k \hat{O}_k + \hat{X}. \tag{53}$$

The coefficients c_i can be found by essentially the same procedure as given by eq. (52). $\hat{X}$ will vanish if a set $\{\hat{O}_i\}$ is found which can represent H exactly.

In general it is not possible to find a such set, else the nuclear eigenvalue problem would be a much simpler one. The next best thing to do is to subdivide the space according to some group G and choose $\{\hat{O}_i\}$ such that a trace-equivalent Hamiltonian, H_{tr}, defined by

$$H_{\mathrm{tr}} = c_1 \hat{O}_1 + c_2 \hat{O}_2 + \cdots + c_k \hat{O}_k, \tag{54}$$

reproduces all the subspace centroids correctly,

$$\langle H \rangle^{(G)} = \langle H_{\mathrm{tr}} \rangle^{(G)}. \tag{55}$$

This is an extension of the concept of trace-equivalent Hamiltonian used in earlier chapters.

By definition, $\hat{X}$, the difference between H and $H_{\rm tr}$, does not contribute to the trace of H although it contributes to $\langle H^2 \rangle$ and the higher-order moments of the density distribution. On the other hand, if a subspace contains a sufficient number of degrees of freedom, we expect that the central limit theorem applies. The distribution is then essentially Gaussian, and the moments higher than the variance are not essential to describe the distribution. Hence, in the limit that the central limit theorem applies, we need only be concerned with the effect of $\hat{X}$ on the width σ of the distribution.

In general, $\sigma_{\rm tr}$, the width of the strength distribution obtained with $H_{\rm tr}$, is narrower than σ obtained with H. If the symmetry G is well chosen, a large fraction of the spectral variance will be contained in the square of $H_{\rm tr}$. The missing part may be taken care of in an approximate way by *renormalizing* $\sigma_{\rm tr}$ so that the total width of the distribution calculated with $H_{\rm tr}$ is the same as with the original Hamiltonian.

Another interesting use of $H_{\rm tr}$ is in obtaining the approximate eigenvalues. When the true Hamiltonian is approximated by $H_{\rm tr}$, expressed in terms of the Casimir operators of a chain of groups, the *trace-equivalent eigenvalues* $E_i^{\rm tr}$ can be written down analytically without having to construct and diagonalize the matrix numerically.

The values of $E_i^{\rm tr}$ must however be *renormalized* for the missing variance. Since the width is the "energy unit" for the distribution, the eigenvalues calculated with $H_{\rm tr}$ are measured in units of $\sigma_{\rm tr}$ while the eigenvalues of H are measured in units of σ. Let

$$r = \frac{\sigma}{\sigma_{\rm tr}} \tag{56}$$

be the ratio between the two widths. Since the centroid of the eigenvalue distribution is preserved when H is approximated by $H_{\rm tr}$, we can restore the width of the distribution of the approximated eigenvalues using the relation

$$E_i^{\rm rn} = C + r(E_i^{\rm tr} - C). \tag{57}$$

Since the loss of distribution width is the dominant effect in the approximation, we expect the values of $E_i^{\rm rn}$ to be close to eigenvalues E_i. Fairly good results in energy positions as well as transition rates compared with full shell model calculations have been obtained by this method for test cases in the ds-shell (Countee *et al.*, 1981).

X.9   Truncation and renormalization of shell model matrices

In many shell model calculations, the dominant part of the low-lying states comes from a small subset of the complete set of basis states. Consequently most of the components of the eigenvectors of interest are so small that they can be ignored without unduly affecting the expectation values and transition rates to be calculated with these eigenvectors. If this is true, it is possible to perform the shell model calculation from the start using only a small part of the full active space.

Truncation of the shell model matrix is usually carried out in terms of single-particle states: however, this is not always the most efficient method to use. Very often we are faced with the dilemma that the elimination of one more single-particle state will lead to too great a loss of accuracy whereas including it will result in too large a matrix. It is clear that a more refined criterion based on the many-particle basis states is desirable. The simplest choice is to use the size of the diagonal matrix elements, or simpler still, to use the contribution from the one-body part of the Hamiltonian to the diagonal matrix elements. However, this is inadequate: the importance of a basis state should be based on its contribution to the region of interest and hence depends not only on the centroid but also on the distribution of the strength of the basis state in question. Statistical spectroscopy is well suited to provide a more detailed condition in terms of subspaces for truncation of the many-particle space. The necessary information can be obtained, $e.g.$, from configuration averaging, or from fixed-JT averaging.

Let us assume that we are interested in all the eigenstates from the ground state up to some energy E'. The contribution of subspace s in this region is given by the integral

$$F_s(E') = \int_{-\infty}^{E'} I_s(E) \, dE \simeq \frac{d_s}{\sigma_s \sqrt{2\pi}} \int_{-\infty}^{E'} \exp\left(-\frac{(E - C_s)^2}{2\sigma_s^2}\right) dE$$

$$= \frac{d_s}{2}\left\{1 + \text{erf}\left(\frac{E - C_s}{\sigma_s \sqrt{2}}\right)\right\}, \tag{58}$$

where a Gaussian approximation is used for the subspace density distribution. The relative contribution of different subspaces is given by

$$S_s(E') = \frac{F_s(E')}{\sum_i F_i(E')}, \tag{59}$$

where the summation in the denominator is over all the subspaces contributing at energy E'. Subspaces with $S_s(E')$ less than some small value can then be discarded since they do not contribute significantly in the region of interest. In this way a realistic estimate can be made for retaining or discarding a group of states in the m-particle space.

The major source of truncation error is in the loss of external width. The total variance σ_s^2 of a subspace s in the full space can be separated into two parts,

$$\sigma_s^2 = \sigma_{sP}^2 + \sigma_{sQ}^2, \tag{60}$$

where σ_{sP}^2 is the variance connecting all the states in the retained, or P space, and σ_{sQ}^2 is the variance connecting s to all the states in the discarded, or Q space. The external variance, σ_{sQ}^2, is therefore given by the sum of the squares of the off-diagonal matrix elements between s and Q. One of the major effects of truncation is the loss of σ_{sQ}^2: after truncation, the variance of the distribution of configuration s is σ_{sP}^2 instead of σ_s^2. Since the purpose of renormalization is to recover from the effect of truncation, we can achieve an important part by scaling the width.

If the central limit theorem applies, the shapes of the distributions are Gaussian both before and after the truncation. The only difference is the width, which measures the spread of the strength distribution. The strength in subspace s after truncation is measured in terms of σ_{sP} instead of σ_s. Since $\sigma_{sP} \leq \sigma_s$, we can recover the effect of the missing σ_{sQ} by renormalizing all the matrix elements in s by the factor

$$r_s = \frac{\sigma_s}{\sigma_{sP}}. \tag{61}$$

In general, r_s is different for different subspaces: the renormalization must therefore be applied to the truncated Hamiltonian matrix before diagonalization. For a typical matrix element H_{ij}, where i belongs to subspace s and j to s', we cannot recover the lost widths in both subspaces at the same time. Since in general $r_s \neq r_{s'}$, the best we can do is to let

$$H_{ij}(R) = H_{ij}\sqrt{r_s r_{s'}}, \tag{62}$$

a geometric mean between the two configurations. Since the truncation and renormalization procedure used here is based on matrices rather than operators, it is different from the usual method used in effective interaction studies to be described in the next section.

Truncation of a matrix is often needed in shell model calculations in order to carry out an otherwise impossible calculation. However, except for truncation according to major shells, renormalization is rarely applied. The main reason for this is the lack of a convenient method rather than of the need. The statistical spectroscopy approach, given more fully in Wong (1978), is both convenient in practice and physically reasonable.

The same renormalization procedure can also be applied to the transition matrix in the truncated space. Here the aim is to recover the norm $\langle \hat{O}^\dagger \hat{O} \rangle$ of the excitation operator $\hat{O}$. In analogy with eq. (60), we have

$$\langle \hat{O}^\dagger \hat{O} \rangle^s = \langle \hat{O}^\dagger \hat{O} \rangle^{s(P)} + \langle \hat{O}^\dagger \hat{O} \rangle^{s(Q)}, \tag{63}$$

where the norm of $\hat{O}$ in subspace s is divided into two parts according to whether the final states belong to the P space or the Q space, as carried out earlier in eq. (60). Since $\langle \hat{O}^\dagger \hat{O} \rangle^{s(Q)}$ leads to final states that are outside the retained space after truncation, it is lost in a calculation carried out in the truncated space. We can recover it approximately by renormalizing the transition matrix elements belonging to subspace s by the ratio

$\langle \hat{O}^\dagger \hat{O} \rangle^s / \langle \hat{O}^\dagger \hat{O} \rangle^{s(P)}$. The accuracy of this procedure is poor if the ratio is much greater than unity since this implies that we have discarded an important part of the space as far as the transition strength is concerned. We should therefore go back and reexamine the truncation procedure and take also the transition strengths into consideration in the selection of the retained space. In this way, we can truncate and renormalize shell model matrices according to both energy and transition strength considerations.

X.10   Effective interaction

For many applications it is more instructive to write the nuclear Hamiltonian in the form

$$H = H_0 + V, \tag{64}$$

where H_0 includes the kinetic energy as well as the average field of all the nucleons on a single particle, and V is the residual interaction. Because of mathematical convenience, H_0 often taken to be an harmonic oscillator Hamiltonian so that the harmonic oscillator wave functions can be used as the single-particle radial wave functions.

The number of single-particle basis states ϕ_i, and hence the number of many-particle basis states Φ, in such a space is in principle infinite. As mentioned in Chapter I, for practical calculations it is necessary to truncate the Hilbert space to a small finite size, say of dimension d. Let us define an operator P which projects out the d basis states to be included in the active space, and let

$$Q = 1 - P$$

be the operator which projects out the states to be ignored. Instead of solving the eigenvalue problem,

$$H \Psi_i = E_i \Psi_i, \tag{65}$$

in the complete Hilbert space, we wish to solve it approximately in the P-space alone using an effective Hamiltonian, H_{eff}, such that the lowest few eigenvalues can be obtained without loss of accuracy,

$$H_{\text{eff}} P \Psi_i = E_i P \Psi_i. \tag{66}$$

The effective interaction $\mathcal{V}_{\text{eff}}$ is then defined by

$$H_{\text{eff}} = H_0 + \mathcal{V}_{\text{eff}}, \tag{67}$$

in analogy with eq. (64).

Since $P + Q = 1$, $P^2 = P$, and $Q^2 = Q$, we obtain from eq. (65) the following set of equations,

$$PHP\Psi_i + PHQ\Psi_i = EP\Psi_i,\qquad\qquad(68a)$$

$$QHP\Psi_i + QHQ\Psi_i = EQ\Psi_i.\qquad\qquad(68b)$$

From eq. (68b) we get the formal solution,

$$Q\Psi_i = \frac{1}{E - QHQ}QHP\Psi_i.$$

On substituting this result into eq. (68a), we obtain

$$PHP\Psi_i + PHQ\frac{1}{E - QHQ}QHP\Psi_i = EP\Psi_i.$$

Using the fact that P and H_0 commute, and that $PQ = QP = 0$, we can rewrite the above equation in the form

$$P\left\{H_0 + V + VQ\frac{1}{E - QHQ}QV\right\}P\Psi_i = EP\Psi_i.$$

Comparing this result with eq. (67), we obtain the effective interaction,

$$\mathcal{V}_{\text{eff}} \equiv V + VQ\frac{1}{E - QHQ}QV = V + VQ\frac{1}{E - Q(H_0 + V)Q}QV.\qquad(69)$$

This interaction is therefore energy dependent, and is usually evaluated in terms of a perturbation series expansion,

$$\mathcal{V}_{\text{eff}} = V + V\frac{Q}{E - H_0}V + V\frac{Q}{E - H_0}V\frac{Q}{E - H_0}V + \cdots$$

$$= \sum_{n=1}^{\infty} V\left(\frac{Q}{(E - H_0)}V\right)^{n-1},\qquad(70)$$

which can be carried to as high an order as feasible.

Two applications of the statistical approach have been made in effective interaction studies. The first one is in checking the convergence of the perturbation series. Traditionally, an exact calculation of eq. (70) beyond the third-order terms ($n = 3$) becomes too complicated to be carried out in full. On the other hand, it is not clear from the known results that the series up to the third order is convergent. Goode and Koltun (1975) found that, whereas an exact calculation was prohibitive, the average of the fourth-order terms over the states with all possible J and T values was relatively easy to obtain. In this way they conclude that the fourth-order terms were as important as the second and third order ones. The convergence of the series is therefore doubtful.

A second application is to evaluate the higher-order terms. Each of the terms in eq. (70), apart from an energy dependent denominator, has the form $V_{PQ}V_{QQ}\cdots V_{QP}$ where all the intermediate states are in the Q space. In terms of single-particle creation and annihilation operators, this part of the terms can be represented by $a_P^\dagger a_P^\dagger a_Q a_Q \cdots a_Q^\dagger a_Q^\dagger a_P a_P$ where the operators in the middle represented by $\cdots$ are exclusively in the Q space.

The general interest is in the two-body matrix elements of $\mathcal{V}_{\text{eff}}$ in the P space. A two-particle state in the P space, with one particle in orbit r and another one in orbit s, angular-momentum coupled to spin-isospin Γ, can be written in the complete Hilbert space as the product of a P space part $|rs\Gamma\rangle$ and a core part $|c\rangle$. The core configuration is usually omitted if the calculation is to be carried out in the P space alone. Let us define a two-particle-state creation operator $Z^\Gamma(rs)$ by

$$|rs\Gamma\rangle \times |c\rangle = \varsigma_{rs}\left(A^r \times A^s\right)^\Gamma |c\rangle \equiv Z^\Gamma(rs)\,|c\rangle, \qquad (71)$$

where the factor ς_{rs} introduced in eq. (IV-12) is included to ensure that the two-particle state created is properly normalized. The hermitian conjugate of $Z^\Gamma(rs)$, represented by $\overline{Z}^\Gamma(rs)$, annihilates a two-particle state.

In terms of the expansion given in eq. (70), the two-body matrix elements of $\mathcal{V}_{\text{eff}}$ in the P space can be written as

$$\mathcal{V}_{rstu}^\Gamma \equiv \langle rs\Gamma|\mathcal{V}_{\text{eff}}|tu\Gamma\rangle = \sum_{n=1}^\infty \langle rs\Gamma|V\left(\frac{Q}{E-H_0}V\right)^{n-1}|tu\Gamma\rangle$$

$$= \sum_{n=1}^\infty \langle rs\Gamma|\mathcal{V}^{(n)}|tu\Gamma\rangle. \qquad (72)$$

In the complete Hilbert space we can write each of the terms in the sum in eq. (72) as the expectation value over the core state $|c\rangle$ using the two-particle creation and annihilation operators defined above,

$$\langle c| \times \langle rs\Gamma|\mathcal{V}^{(n)}|tu\Gamma\rangle \times |c\rangle = \langle c|\left(\overline{Z}^\Gamma(rs) \times \mathcal{V}^{(n)} \times Z^\Gamma(tu)\right)^0|c\rangle$$

$$\equiv \langle c|\hat{O}^{(n)}|c\rangle, \qquad (73)$$

where we have defined

$$\hat{O}^{(n)} \equiv \left(\overline{Z}^\Gamma(rs) \times \mathcal{V}^{(n)} \times Z^\Gamma(tu)\right)^0,$$

for convenience in the discussion below.

Since the dimension of the core configuration is unity, the expectation value and the average trace are equal, and we have

$$\langle c | \hat{O}^{(n)} | c \rangle = \langle \hat{O}^{(n)} \rangle^{(c)} \, ,$$

where c on the right hand side denotes the configuration with the core orbits filled and the other orbits empty. As a result, all the techniques developed for configuration averaging can be used to calculate the various terms in the perturbation series of eq. (70).

There are two advantages in using configuration averaging. First, the method can handle $\hat{O}^{(n)}$ in the fully JT-coupled form. That is, all the single-particle creation and annihilation operators constituting $O^{(n)}$ are angular-momentum coupled to definite spin and isospin. This is far simpler in general than in the M-scheme. Second, in configuration averaging, the tedious procedure of finding the basic diagrams and reducing the diagrams into algebraic expressions can be carried out on a computer. As a result, one can afford to try different ways of carrying out the reductions to arrive at the simplest possible expressions. For example, all twenty-six diagrams in the fourth order of the perturbation series have been evaluated by Chang, Vincent, and Wong (1984) in this way. The resulting expressions have been compared with those given by Barrett and Kirson (1970), and in several cases simpler, therefore faster in calculation, expressions are found. Perhaps of more importance, the configuration averaging technique can produce the algebraic expressions for even higher-order terms in the perturbation series in a relatively easy way.

Most of these and other applications unique to statistical spectroscopy are still in their infancy. Many more development studies and applications are required before they can become standard procedures in the study of nuclear systems. The potential of statistical spectroscopy is far from being exhausted by the examples described here. New and different uses are expected to be forthcoming.

Appendix A

Some Useful Combinatorial Identities

We give here some of the relations between binomial coefficients used in the text. In general they are simple identities; other formulae and more rigorous proofs can be found in Riordan (1968).

For m and n integers, a binomial coefficient can be defined as

$$\binom{n}{m} = \begin{cases} \frac{n!}{m!(n-m)!} & \text{if } n \geq m, \\ 0 & \text{otherwise.} \end{cases} \tag{1}$$

From the definition, it is clear that

$$\binom{n}{m} = \binom{n}{n-m}, \tag{2}$$

and

$$\binom{n}{0} = 1, \qquad \binom{0}{m} = \delta_{m0}. \tag{3}$$

Similarly, for $m > 0$,

$$\binom{n}{-m} = 0, \tag{4}$$

and

$$\binom{-n}{m} = (-1)^m \binom{n+m-1}{m}, \tag{5}$$

for both $m > 0$ and $n > 0$.

The recurrence relation

$$\binom{n}{m} = \binom{n-1}{m} + \binom{n-1}{k-1}, \tag{6}$$

can be derived in the following way using eq. (1),

$$\binom{n-1}{m} + \binom{n-1}{m-1} = \frac{(n-1)!}{(n-m-1)!\,m!} + \frac{(n-1)!}{(n-m)!\,(m-1)!}$$

$$= \frac{(n-1)!}{(n-m)!\,m!}\{(n-m)+m\}$$

$$= \frac{n(n-1)!}{(n-m)!\,m!} = \binom{n}{m}.$$

Similarly, by rearranging the factorials, we can show that

$$
\binom{n}{m}\binom{m}{p} = \frac{n!}{(n-m)!\,m!}\frac{m!}{(m-p)!\,p!}
$$

$$
= \frac{n!}{(n-p)!\,p!}\frac{(n-p)!}{(n-m)!\,(m-p)!} = \binom{n}{p}\binom{n-p}{m-p} \tag{7}
$$

$$
= \frac{n!}{(m-p)!\,(n-m+p)!}\frac{(n-m+p)!}{(n-m)!\,p!}
$$

$$
= \binom{n}{m-p}\binom{m-n+p}{p}. \tag{8}
$$

These relations are useful to prove other identities.

Using the binomial series,

$$
(1+x)^n = \sum_{r=0}^{n}\binom{n}{r}x^r, \tag{9}
$$

we obtain, by taking $x = +1$,

$$
\sum_{r=0}^{n}\binom{n}{r} = 2^n, \tag{10}
$$

and by taking $x = -1$,

$$
\sum_{r=0}^{n}(-1)^r\binom{n}{r} = (1-1)^n = \delta_{n0}. \tag{11}
$$

From eq. (9), we have for any integer n,

$$
(1+x)^n = (1+x)^{n-m}(1+x)^m = \sum_{r=0}^{n}\sum_{s=0}^{r}\binom{n-m}{s}\binom{m}{r-s}x^r.
$$

This gives us the relation

$$
\binom{n}{r} = \sum_{s=0}^{r}\binom{n-m}{s}\binom{m}{r-s}, \tag{12}
$$

used in several places in the text.

For $n \geq m$, we can derive the following orthogonality relation

$$
\sum_{r=m}^{n}(-1)^r\binom{n}{r}\binom{r}{m} = \sum_{r=m}^{n}(-1)^r\binom{n}{m}\binom{n-m}{r-m}
$$

$$
= (-1)^m\binom{n}{m}\sum_{s=0}^{n-m}(-1)^s\binom{n-m}{s}
$$

$$
= (-1)^m\binom{n}{m}\delta_{nm} = (-1)^m\delta_{nm}, \tag{13}
$$

by using eqs. (7) and (11).

The Vandermonde convolution formula,

$$\binom{n}{m} = \sum_{r=0} \binom{n-p}{m-r}\binom{p}{r},$$

(14)

was derived in Chapter IV. An extension of it,

$$\binom{n-p}{m} = \sum_r (-1)^r \binom{p}{r}\binom{n-r}{m-r},$$

(15)

can be obtained as in eq. (5) of Chapter 1 in Riordan (1968). Using first eq. (5) and then eq. (13), we obtain

$$\binom{n-p}{m} = (-1)^m \binom{p-n+m-1}{m}$$

$$= (-1)^m \sum_r \binom{p}{m-r}\binom{-n+m-1}{r}$$

$$= (-1)^m \sum_r \binom{p}{m-r}(-1)^r \binom{n-m+r}{r}$$

$$= \sum_r (-1)^r \binom{p}{r}\binom{n-r}{m-r},$$

(16)

where the third line is obtained by using eq. (5) again, and the last line by changing the summation index from r to $(m-r)$. By substituting $(N-t)$ for n, $(m-t)$ for m, and $(p-t)$ for p, we obtain from eq. (16) the result

$$\binom{N-p}{m-t} = \sum_r (-1)^r \binom{p-t}{r}\binom{N-t-r}{m-t-r}$$

$$= \sum_r (-1)^r \binom{p-t}{p-t-r}\binom{N-t-r}{m-t-r}$$

$$= \sum_k (-1)^{k-t} \binom{p-t}{p-k}\binom{N-k}{m-k},$$

(17)

which was used to prove eq. (V-20).

We shall prove here an identity used in deriving eq. (IV-18),

$$\sum_{s=k}^{q} (-1)^{k+s} \binom{m}{s}\binom{s}{k} = \binom{m}{k}\binom{q-m}{q-k}.$$

(18)

Using eq. (7), we can rewrite the sum on the left hand side as

$$\sum_{t=k}^{r}(-1)^{k+t}\binom{m}{t}\binom{t}{k}=\sum_{t=k}^{r}(-1)^{k+t}\binom{m}{k}\binom{m-k}{t-k}$$

$$=\binom{m}{k}\sum_{u=0}^{r-k}(-1)^{u}\binom{m-k}{u}. \qquad (19)$$

The summation over u cannot be carried out easily since the upper limit is $(r-k)$, whereas all the sum rules derived above have natural upper limits imposed by the zeros or the *boundary conditions* of the binomial coefficients involved. We can make use of the Vandermonde relation to overcome this problem. On putting $n = m$ in eq. (15), and replacing n by $(r - k)$ and p by $(m - k)$, we obtain

$$\binom{r-m}{r-k}=\sum_{u=0}(-1)^{u}\binom{r-k-u}{r-k-u}\binom{m-k}{u}. \qquad (20)$$

Since $\binom{m}{n} = 0$ for $m < 0$, the upper limit of the summation above is now restricted to $(r - k)$ by the boundary condition of the first binomial coefficient. Furthermore, since $\binom{n}{n} = 1$, we have

$$\binom{r-m}{r-k}=\sum_{u=0}^{r-k}(-1)^{u}\binom{m-k}{u}. \qquad (21)$$

On substituting this result into eq. (19), we obtain the identity in eq. (18).

Appendix B

Angular Momentum Recoupling Coefficients

A Clebsch-Gordan coefficient $\langle rpsq|tm \rangle$ is defined in terms of the coupling of two spherical tensors with ranks r and s to a product tensor of rank t:

$$(T^r \times U^s)_m^t = \sum_{pq} \langle rpsq|tm \rangle T_p^r U_q^s. \tag{1}$$

Often it is more convenient to write the coupling coefficent in terms of a $3j$-symbol related to a Clebsch-Gordan coefficient through

$$\begin{pmatrix} j_1 & j_2 & j_3 \\ m_1 & m_2 & m_3 \end{pmatrix} = \frac{(-1)^{j_1 - j_2 - m_3}}{\sqrt{2j_3 + 1}} \langle j_1 m_1 j_2 m_2 | j_3 \ -m_3 \rangle. \tag{2}$$

The symmetry and orthogonality properties of the Clebsch-Gordan coefficients can be best displayed in terms of $3j$-symbols:

$$\begin{pmatrix} j_1 & j_2 & j_3 \\ m_1 & m_2 & m_3 \end{pmatrix} = \begin{pmatrix} j_2 & j_3 & j_1 \\ m_2 & m_3 & m_1 \end{pmatrix} = \begin{pmatrix} j_3 & j_1 & j_2 \\ m_3 & m_1 & m_2 \end{pmatrix}$$

$$= (-1)^{j_1 + j_2 + j_3} \begin{pmatrix} j_1 & j_3 & j_2 \\ m_1 & m_3 & m_2 \end{pmatrix}$$

$$= (-1)^{j_1 + j_2 + j_3} \begin{pmatrix} j_1 & j_2 & j_3 \\ -m_1 & -m_2 & -m_3 \end{pmatrix}. \tag{3}$$

To simplify the notation we shall use $\Delta(j_1 j_2 j_3)$ to stand for the triangular relation between tensors j_1, j_2, and j_3 such that

$$\Delta(j_1 j_2 j_3) = \begin{cases} 1 & \text{for} \quad j_1 + j_2 \geq j_3 \geq |j_1 - j_2|, \\ 0 & \text{otherwise.} \end{cases}$$

The orthogonality relations can be written in the form

$$\sum_{m_1 m_2} \begin{pmatrix} j_1 & j_2 & j_3 \\ m_1 & m_2 & m_3 \end{pmatrix} \begin{pmatrix} j_1 & j_2 & j_3' \\ m_1 & m_2 & m_3' \end{pmatrix} = \Delta(j_1 j_2 j_3) \frac{\delta_{j_3 j_3'} \delta_{m_3 m_3'}}{[j_3]},$$

$$\sum_{m_1 m_2 m_3} \begin{pmatrix} j_1 & j_2 & j_3 \\ m_1 & m_2 & m_3 \end{pmatrix} \begin{pmatrix} j_1 & j_2 & j_3 \\ m_1 & m_2 & m_3 \end{pmatrix} = \Delta(j_1 j_2 j_3),$$

$$\sum_{j_3 m_3} [j_3] \begin{pmatrix} j_1 & j_2 & j_3 \\ m_1 & m_2 & m_3 \end{pmatrix} \begin{pmatrix} j_1 & j_2 & j_3 \\ m_1' & m_2' & m_3 \end{pmatrix} = \Delta(j_1 j_2 j_3) \, \delta_{m_1 m_1'} \delta_{m_2 m_2'}, \tag{4}$$

where $[j_3] \equiv (2j_3 + 1)$.

A Racah coefficient $W(j_1 j_2 J j_3; J_{12} J_{23})$ is used to recouple three spherical tensors from the form $(((j_1 \times j_2)^{J_{12}} \times j_3)^{J_{23}})^J$ to the form $(j_1 \times (j_2 \times j_3)^{J_{23}})^J$,

$$((j_1 \times j_2)^{J_{12}} \times j_3)^J = \sum_{J_{23}} [J_{12} J_{23}]^{\frac{1}{2}} W(j_1 j_2 J j_3; J_{12} J_{23})$$

$$\times (j_1 \times (j_2 \times j_3)^{J_{23}})^J. \qquad (5)$$

In the same way as for Clebsch-Gordan coefficients, it is more convenient to display the symmetry properties of Racah coefficients in terms of $6j$-symbols defined by

$$\begin{Bmatrix} j_1 & j_2 & J_{12} \\ j_3 & J & J_{23} \end{Bmatrix} = (-1)^{j_1+j_2+j_3+J} W(j_1 j_2 J j_3; J_{12} J_{23}), \qquad (6)$$

and they are

$$\begin{Bmatrix} j_1 & j_2 & j_3 \\ j_4 & j_5 & j_6 \end{Bmatrix} = \begin{Bmatrix} j_2 & j_3 & j_1 \\ j_5 & j_6 & j_4 \end{Bmatrix} = \begin{Bmatrix} j_3 & j_1 & j_2 \\ j_6 & j_4 & j_5 \end{Bmatrix} = \begin{Bmatrix} j_2 & j_1 & j_3 \\ j_5 & j_4 & j_6 \end{Bmatrix} = \begin{Bmatrix} j_4 & j_5 & j_3 \\ j_1 & j_2 & j_6 \end{Bmatrix}. \qquad (7)$$

A $6j$-symbol is related to a product of four $3j$-symbols in the form

$$\begin{pmatrix} j_1 & j_2 & j_3 \\ m_1 & m_2 & m_3 \end{pmatrix} \begin{Bmatrix} j_1 & j_2 & j_3 \\ l_1 & l_2 & l_3 \end{Bmatrix} = \sum_{m'_1 m'_2 m'_3} (-1)^{l_1+l_2+l_3+m'_1+m'_2+m'_3}$$

$$\times \begin{pmatrix} j_1 & l_2 & l_3 \\ m_1 & m'_2 & -m'_3 \end{pmatrix} \begin{pmatrix} l_1 & j_2 & l_3 \\ -m'_1 & m_2 & m'_3 \end{pmatrix} \begin{pmatrix} l_1 & l_2 & j_3 \\ m'_1 & -m'_2 & m_3 \end{pmatrix}. \qquad (8)$$

From this, we can obtain an expression for the $6j$-symbol in terms of a sum over products of four $3j$-symbols.

The following special values and sum rules are often useful. Given three tensors, a, b, and c forming a triangle, *i.e.*, $\Delta(abc) = 1$, we have

$$\sum_c [c] = [a, b] \sum_c [c] \begin{Bmatrix} a & a & 0 \\ b & b & c \end{Bmatrix} \begin{Bmatrix} a & a & 0 \\ b & b & c \end{Bmatrix} = [a, b]. \qquad (9)$$

Furthermore, for $b = a$,

$$\sum_c (-1)^c [c] = [a, a] \sum_c (-1)^c [c] \begin{Bmatrix} a & a & 0 \\ a & a & c \end{Bmatrix} \begin{Bmatrix} a & a & 0 \\ a & a & c \end{Bmatrix} = (-1)^{2a}[a]. \qquad (10)$$

If one of the six arguments of the $6j$-symbol is zero,

$$\begin{Bmatrix} j_1 & j'_1 & 0 \\ j_2 & j'_2 & j \end{Bmatrix} = (-1)^{j_1+j_2+j_3} \frac{\delta_{j_1 j'_1} \delta_{j_2 j'_2}}{[j_1 j_2]^{\frac{1}{2}}}. \qquad (11)$$

There are two sum rules involving a single $6j$-symbol,

$$\sum_j (-1)^{j_1+j_2+j} [j] \begin{Bmatrix} j_1 & j_1 & j' \\ j_2 & j_2 & j \end{Bmatrix} = [j_1 j_2]^{\frac{1}{2}} \delta_{j'0} , \qquad (12)$$

and

$$\sum_j (-1)^{2j} [j] \begin{Bmatrix} j_1 & j_2 & j \\ j_1 & j_2 & j' \end{Bmatrix} = 1. \qquad (13)$$

An alternative form of the above can be written as

$$\sum_j [j] \begin{Bmatrix} j_1 & j_2 & j \\ j_1 & j_2 & j' \end{Bmatrix} = (-1)^{2(j_1+j_2)}. \qquad (13a)$$

For the product of two $6j$-symbols we have the following two sum rules,

$$\sum_j (-1)^{j+j'+j''} [j] \begin{Bmatrix} j_1 & j_2 & j' \\ j_3 & j_4 & j \end{Bmatrix} \begin{Bmatrix} j_1 & j_3 & j'' \\ j_2 & j_4 & j \end{Bmatrix} = \begin{Bmatrix} j_1 & j_2 & j' \\ j_4 & j_3 & j'' \end{Bmatrix} , \qquad (14)$$

and

$$\sum_j (-1)^{2j} [j] \begin{Bmatrix} j_1 & j_2 & j' \\ j_3 & j_4 & j \end{Bmatrix} \begin{Bmatrix} j_1 & j_2 & j'' \\ j_3 & j_4 & j \end{Bmatrix} = (-1)^{2(j_1+j_4)} \frac{\delta_{j'j''}}{[j']} . \qquad (15)$$

An alternative form of eq. (15) is

$$\sum_j [j] \begin{Bmatrix} j_1 & j_2 & j' \\ j_3 & j_4 & j \end{Bmatrix} \begin{Bmatrix} j_1 & j_2 & j'' \\ j_3 & j_4 & j \end{Bmatrix} = \frac{\delta_{j'j''}}{[j']} . \qquad (15a)$$

With three $6j$-symbols we have

$$\sum_k (-1)^{k_1+k_2+k} [k] \begin{Bmatrix} k_1 & k_2 & k \\ j_1' & j_1 & j_2'' \end{Bmatrix} \begin{Bmatrix} k_1 & k_2 & k \\ j_2' & j_2 & j_1'' \end{Bmatrix} \begin{Bmatrix} j_1 & j_1' & k \\ j_2' & j_2 & j \end{Bmatrix}$$

$$= (-1)^{j_1+j_2+j_1'+j_2'+j_1''+j_2''+j} \begin{Bmatrix} j_1 & j_2 & j \\ j_1'' & j_2'' & k_1 \end{Bmatrix} \begin{Bmatrix} j_1' & j_2' & j \\ j_1'' & j_2'' & k_2 \end{Bmatrix} . \qquad (16)$$

Summation of a product of three $6j$-symbols in terms of a $9j$-symbol is given below in eqs. (19).

A $9j$-symbol, defined by

$$\left((j_1 \times j_2)^{J_{12}} \times (j_3 \times j_4)^{J_{34}} \right)^J$$

$$= \sum_{J_{13} J_{24}} [J_{12} J_{34} J_{13} J_{24}]^{\frac{1}{2}} \begin{Bmatrix} j_1 & j_2 & J_{12} \\ j_3 & j_4 & J_{34} \\ J_{13} & J_{24} & J \end{Bmatrix} \left((j_1 \times j_3)^{J_{13}} \times (j_2 \times j_4)^{J_{24}} \right)^J ,$$

$$\qquad (17)$$

is used to recouple four spherical tensors from the form $((j_1 \times j_2)^{J_{12}} \times (j_3 \times j_4)^{J_{34}})^J$ to the form $((j_1 \times j_3)^{J_{13}} \times (j_2 \times j_4)^{J_{24}})^J$ for j_2 commuting with j_3. It is symmetrical under a rotation of the arguments in the following ways,

$$
\begin{Bmatrix} j_1 & j_2 & j_3 \\ j_4 & j_5 & j_6 \\ j_7 & j_8 & j_9 \end{Bmatrix}
= \begin{Bmatrix} j_1 & j_4 & j_7 \\ j_2 & j_5 & j_8 \\ j_3 & j_6 & j_9 \end{Bmatrix}
= \begin{Bmatrix} j_7 & j_8 & j_9 \\ j_1 & j_2 & j_3 \\ j_4 & j_5 & j_6 \end{Bmatrix}
= \begin{Bmatrix} j_4 & j_5 & j_6 \\ j_7 & j_8 & j_9 \\ j_1 & j_2 & j_3 \end{Bmatrix}
$$

$$
= (-1)^{j_1+j_2+j_3+j_4+j_5+j_6+j_7+j_8+j_9}
\begin{Bmatrix} j_4 & j_5 & j_6 \\ j_1 & j_2 & j_3 \\ j_7 & j_8 & j_9 \end{Bmatrix}. \tag{18}
$$

A $9j$-symbol can be written as a sum of products of three $6j$-symbols,

$$
\begin{Bmatrix} j_1 & j_2 & J_{12} \\ j_3 & j_4 & J_{34} \\ J_{13} & J_{24} & J \end{Bmatrix}
= \sum_{J'} (-1)^{2J'} [J'] \begin{Bmatrix} j_1 & J & J' \\ J_{34} & j_2 & J_{12} \end{Bmatrix}
$$

$$
\times \begin{Bmatrix} j_1 & J & J' \\ J_{24} & j_3 & J_{13} \end{Bmatrix}
\begin{Bmatrix} J_{34} & j_2 & J' \\ J_{24} & j_3 & j_4 \end{Bmatrix}, \tag{19}
$$

or alternatively,

$$
\begin{Bmatrix} j_1 & j_2 & J_{12} \\ j_3 & j_4 & J_{34} \\ J_{13} & J_{24} & J \end{Bmatrix}
= (-1)^{2(j_1+J)} \sum_{J'} [J'] \begin{Bmatrix} j_1 & J & J' \\ J_{34} & j_2 & J_{12} \end{Bmatrix}
$$

$$
\times \begin{Bmatrix} j_1 & J & J' \\ J_{24} & j_3 & J_{13} \end{Bmatrix}
\begin{Bmatrix} J_{34} & j_2 & J' \\ J_{24} & j_3 & j_4 \end{Bmatrix}. \tag{19a}
$$

This expression can also be written in terms of a sum rule

$$
\sum_J [J] \begin{Bmatrix} J_{12} & J_{34} & J \\ J_{13} & J_{24} & J' \end{Bmatrix}
\begin{Bmatrix} j_1 & j_2 & J_{12} \\ j_3 & j_4 & J_{34} \\ J_{13} & J_{24} & J \end{Bmatrix}
$$

$$
= (-1)^{2J'} \begin{Bmatrix} j_1 & j_3 & J_{13} \\ J_{34} & J' & j_4 \end{Bmatrix}
\begin{Bmatrix} j_2 & j_4 & J_{24} \\ J' & J_{12} & j_1 \end{Bmatrix}. \tag{20}
$$

If one of the nine arguments is zero, a $9j$-symbol reduces to a $6j$-symbol:

$$
\begin{Bmatrix} j_1 & j_2 & J_1 \\ j_3 & j_4 & J_1' \\ J_2 & J_2' & 0 \end{Bmatrix}
= \frac{(-1)^{j_2+j_3+J_1+J_2}}{[J_1 J_2]^{\frac{1}{2}}}
\begin{Bmatrix} j_1 & j_2 & J_1 \\ j_4 & j_3 & J_2 \end{Bmatrix} \delta_{J_1,J_1'} \delta_{J_2,J_2'}. \tag{21}
$$

The orthogonality relation between two $9j$-symbols is given by

$$
\sum_{J_{13} J_{24}} [J_{13} J_{24}]
\begin{Bmatrix} j_1 & j_2 & J_{12} \\ j_3 & j_4 & J_{34} \\ J_{13} & J_{24} & J \end{Bmatrix}
\begin{Bmatrix} j_1 & j_2 & J_{12}' \\ j_3 & j_4 & J_{34}' \\ J_{13} & J_{24} & J \end{Bmatrix}
= \frac{\delta_{J_{12} J_{12}'} \delta_{J_{34} J_{34}'}}{[J_{12} J_{34}]}, \tag{22}
$$

and the sum rule involving a product of two $9j$-symbols is given by

$$\sum_{J_{13}J_{24}} (-1)^{j_2+j_4+J_{24}} [J_{13}J_{24}] \left\{ \begin{array}{ccc} j_1 & j_2 & J_{12} \\ j_3 & j_4 & J_{34} \\ J_{13} & J_{24} & J \end{array} \right\} \left\{ \begin{array}{ccc} j_1 & j_4 & J_{14} \\ j_3 & j_2 & J_{23} \\ J_{13} & J_{24} & J \end{array} \right\}$$

$$= (-1)^{(j_3+j_4+J_{34})-(j_2+j_3+J_{23})} \left\{ \begin{array}{ccc} j_1 & j_4 & J_{14} \\ j_2 & j_3 & J_{23} \\ J_{12} & J_{34} & J \end{array} \right\}. \qquad (23)$$

Appendix C

Irreducible Spherical Tensors

An irreducible spherical tensor T_q^k of rank k is a quantity with $(2k+1)$ components distinguished by different values of q, the projection of k on the quantization axis. The possible values of q range from $-k$ to k in integer steps. Under a rotation of the coordinate system, the different components transform into each other according to the relation

$$T_q^{k\prime} = \sum_p T_p^k D_{pq}^k(\alpha\beta\gamma). \tag{1}$$

Here $(\alpha\beta\gamma)$ are the Euler angles which rotate the unprimed system to the primed system, and $D_{pq}^k(\alpha\beta\gamma)$ is the rotation matrix defined according to Brink and Satchler (1968). The components of T_q^k therefore form an irreducible representation of the rotation group.

The different components $Y_m^l(\theta, \phi)$ of a spherical harmonic of order l transform into each other according to the relation

$$Y_m^l(\theta', \phi') = \sum_n Y_n^l(\theta, \phi) D_{nm}^l(\alpha\beta\gamma), \tag{2}$$

when the axes of the unprimed system are rotated by the Euler angles $(\alpha\beta\gamma)$ into the primed system. Integer-rank spherical tensors are therefore closely related to spherical harmonics since they have the same transformation property. Spherical tensors are however more general quantities since they can also take on half-integer ranks of which there is no counter part in spherical harmonics.

Using the Wigner-Eckart theorem, the matrix elements of a spherical tensor operator between states of definite angular momenta can be written as the product of a reduced matrix element and a $3j$-symbol,

$$\langle JM|T_q^k|J'M'\rangle = (-1)^{J-M} \begin{pmatrix} J & k & J' \\ -M & q & M' \end{pmatrix} \langle J\|T^k\|J'\rangle. \tag{3}$$

In terms of a Clebsch-Gordan coefficient instead of $3j$-symbol, eq. (3) can be written in the form

$$\langle JM|T_q^k|J'M'\rangle = (-1)^{2k} \frac{\langle J'M'kq|JM\rangle}{\sqrt{(2J+1)}} \langle J\|T^k\|J'\rangle. \tag{3a}$$

There are several variants of the definition of a reduced matrix element from the above: the phase factor, $(-1)^{2k}$, which is important when we deal with half-integer rank operators, and the factor $\sqrt{(2J+1)}$ are sometimes omitted. A triple bar matrix element, $\langle JT \| T^k \| J'T' \rangle$, is sometimes used to denote a matrix element reduced in both spin and isospin whenever there is a need to distinguish from the double bar matrix elements denoting matrix elements reduced in spin but not isospin.

The conjugate $(T_q^k)^\dagger$ of a tensor operator, defined by

$$\langle JM|(T_q^k)^\dagger|J'M'\rangle = \langle J'M'|T_q^k|JM\rangle^*,$$

is not a proper spherical tensor since it transforms under a rotation according to

$$(T_q^{k'})^\dagger = \sum_p (T_p^k)^\dagger \left(\mathcal{D}_{pq}^k(\alpha\beta\gamma) \right)^*. \tag{4}$$

In order to have the proper tensorial transformation property as given in eq. (1), we define an adjoint tensor $\overline{T}_q^k$ by

$$\overline{T}_q^k \equiv (-1)^{p-q}(T_{-q}^k)^\dagger. \tag{5}$$

Since

$$\mathcal{D}_{rs}^k(\alpha\beta\gamma)^* = (-1)^{r-s}\mathcal{D}_{-r-s}^k(\alpha\beta\gamma),$$

$\overline{T}_q^k$ has the same transformation property as given by eq. (1). The q-dependence in the phase factor of eq. (5) is essential; the value of p, however, is somewhat arbitrary. Both $p = 0$ and $p = k$ are used in the literature and we adopt $p = k$ for reasons stated below.

A hermitian or self-adjoint tensor is defined by

$$T_q^k \equiv \overline{T}_q^k = (-1)^{p-q} \left(T_{-q}^k \right)^\dagger. \tag{6}$$

For spherical harmonics,

$$\left(Y_M^L\right)^\dagger = \left(Y_M^L\right)^* = (-1)^M Y_{-M}^L,$$

and (the spherical components of) spin J,

$$J_q^\dagger = (-1)^q J_{-q},$$

where $J_0 = J_z$ and $J_{\pm 1} = \mp \frac{1}{\sqrt{2}}(J_x \pm iJ_y)$, the convention adopted corresponding to $p = 0$. However this choice is inconvenient in general for two reasons. First, half-integer rank tensors cannot be made self-adjoint by this convention. Second, a product of two self-adjoint tensors is not self-adjoint. This can be shown by the following example. Let T^r and U^s

be two self-adjoint tensors. The m-component of their rank-t product is given by eq. (B-1),

$$(T^r \times U^s)^t_m = \sum_{\mu\nu} \langle r\mu s\nu | tm \rangle \, T^r_\mu U^s_\nu.$$

The adjoint of the product is then

$$\overline{(T^r \times U^s)}^t_m = \sum_{\mu\nu} \langle r\mu s\nu | tm \rangle \, \overline{U}^s_\nu \overline{T}^r_\mu = (-1)^{r+s-t} (\overline{U}^s \times \overline{T}^r)^t_m, \qquad (7)$$

where the final result is arrived at using the properties of Clebsch-Gordan coefficients given in eqs. (B-2,3). In order for the product to be also self-adjoint according to eq. (6), it is necessary that

$$(T^r \times U^s)^t_m = (-1)^{p_t - m} \{(T^r \times U^s)^t_{-m}\}^\dagger$$

$$= (-1)^{p_t - m} \sum_{\mu\nu} \langle r, -\mu, s, -\nu | t, -m \rangle \, (T^r_{-\mu} U^s_{-\nu})^\dagger$$

$$= (-1)^{p_t - m} \sum_{\mu\nu} \langle r, -\mu, s, -\nu | t, -m \rangle \, (U^s_{-\nu})^\dagger (T^r_{-\mu})^\dagger$$

$$= (-1)^{p_t - m} \sum_{\mu\nu} \langle s\nu r\mu | tm \rangle (-1)^{\nu - p_s} \overline{U}^s_\nu (-1)^{\mu - p_r} \overline{T}^r_\mu$$

$$= (-1)^{p_t - p_r - p_s} (\overline{U}^s \times \overline{T}^r)^t_m, \qquad (8)$$

since $\mu + \nu = m$. Comparing eq. (7) with (8), it is clear that in eq. (6) the choice $p = k$, the rank of the tensor, will ensure the product tensor to be also self adjoint. We therefore adopt

$$\overline{T}^k_q \equiv (-1)^{k-q} (T^k_{-q})^\dagger, \qquad (5a)$$

for adjoint tensors and

$$T^k_q \equiv \overline{T}^k_q = (-1)^{k-q} \left(T^k_{-q} \right)^\dagger, \qquad (6a)$$

for self-adjoint tensors.

For hermitian operators, we can derive the following relationship between the reduced matrix elements of T^k_q and $\overline{T}^k_q$. Starting from eq. (5), and making use of the fact that the matrix element vanishes unless $q = M - M'$, we obtain

$$\langle JM | T^k_q | J'M' \rangle = \langle J'M' | (T^k_q)^\dagger | JM \rangle = (-1)^{p+M'-M} \langle J'M' | \overline{T}^k_{-q} | JM \rangle,$$

where $p = k$ if eq. (5a) is adopted. Applying the Wigner-Eckart theorem to both sides and making use of the symmetry relations of $3j$-symbols given by eq. (B-3), we have

$$(-1)^{J-M} \begin{pmatrix} J & k & J' \\ -M & q & M' \end{pmatrix} \langle J \| T^k \| J' \rangle$$

$$= (-1)^{p+M'-M}(-1)^{J'-M'} \begin{pmatrix} J' & k & J \\ -M' & -q & M \end{pmatrix} \langle J \| \overline{T}^k \| J' \rangle$$

$$= (-1)^{p+M'-M}(-1)^{J'-M'} \begin{pmatrix} J & k & J' \\ -M & q & M' \end{pmatrix} \langle J \| \overline{T}^k \| J' \rangle$$

$$= (-1)^{p+J'-J}(-1)^{J-M} \begin{pmatrix} J & k & J' \\ -M & q & M' \end{pmatrix} \langle J \| \overline{T}^k \| J' \rangle.$$

Hence

$$\langle J \| T^k \| J' \rangle = (-1)^{k+J-J'} \langle J' \| \overline{T}^k \| J \rangle, \tag{9}$$

where we have taken $p = k$ according to eq. (6a).

We shall now extend the idea of spherical tensors to second quantized operators. The operator $a_{j,m}$, which annihilates a particle occupying the single-particle state $|jm\rangle$, is the m component of a tensor of rank j. Its conjugate $a_{j,m}^\dagger$ is not a proper spherical tensor because of eq. (4). It is therefore more convenient to define a new set of single-particle creation and annihilation operators,

$$A_m^j \equiv a_{j,m}^\dagger, \qquad \text{and} \qquad B_m^j \equiv (-1)^{j+m} a_{j,-m}, \tag{10}$$

which are proper spherical tensors.

From eq. (5a) it is obvious that

$$A_m^j = \overline{B}_m^j, \tag{11}$$

and

$$\overline{A}_m^j = \overline{\overline{B}_m^j} = (-1)^{2j} B_m^j. \tag{12}$$

For identical particles, j is a half-integer and we have $\overline{A}_m^j = -B_m^j$. However, if the definition of single-particle operators includes also the isospin labels (t, z), i.e., A_{mz}^{jt} creates a particles in the state (j, m, t, z), then

$$\overline{A}_{mz}^{jt} = (-1)^{2(j+t)} B_{mz}^{jt} = B_{mz}^{jt}. \tag{12a}$$

The notation can be simplified if we use a single Greek letter to represent both spin-isospin ranks in the isospin formalism and spin alone in the np-formalism. That is, $\rho \equiv (j, t)$ and $\mu \equiv (m, z)$, or $\rho = j$ and $\mu = m$, as the case may require. Both eqs. (12) and (12a) are represented by

$$\overline{A}_\mu^\rho = (-1)^{2\rho} B_\mu^\rho. \tag{12b}$$

The commutation relation between A^ρ and $B^{\rho'}$ can be obtained from that of $a^\dagger$ and a given in eqs. (V-2). In coupled form, it is expressed as

$$\left(A^\rho \times B^{\rho'}\right)^\Gamma + (-1)^{\rho+\rho'-\Gamma}\left(B^{\rho'} \times A^\rho\right)^\Gamma = [\rho]^{\frac{1}{2}}\delta_{\rho\rho'}\delta_{\Gamma 0}. \tag{13}$$

Except for a normalization factor $[\Delta]^{\frac{1}{2}}$, the angular-momentum coupled product of A^ρ and $B^{\rho'}$ is the Racah unit tensor,

$$u^\Delta_{\rho\rho'} = [\Delta]^{-\frac{1}{2}}(A^\rho \times B^{\rho'})^\Delta. \tag{14}$$

Since Δ is formed from the vector addition of ρ and ρ', we have the following value for the reduced matrix element in single-particle space,

$$\langle\rho''\|(A^\rho \times B^{\rho'})^\Delta\|\rho'''\rangle = [\Delta]^{\frac{1}{2}}\delta_{\rho''\rho}\delta_{\rho'\rho'''},$$

as can be seen by using eqs. (16) and (20) ahead.

By definition, when A^ρ acts on the vacuum $|0\rangle$, it creates a state of one particle:

$$A^\rho|0\rangle = |\rho\rangle.$$

The projection quantum numbers may be omitted from now on since we shall be dealing mostly with reduced matrix elements. When A^ρ acts on an m-particle state, a state of $(m+1)$ particles is produced,

$$\{A^\rho \times |\rho^m\Gamma'x\rangle\}^\Gamma = \alpha_{m+1}|\rho^{m+1}\Gamma y\rangle, \tag{15}$$

where x and y represent all the quantum numbers other than particle number, spin, and isospin. The state produced is in general not normalized. The normalization constant α_{m+1} is complicated by the antisymmetrization requirement and is usually given in terms of coefficients of fractional parentage $\langle(m+1)\Gamma x|m\Gamma'y\rangle$ in the form

$$\langle\rho^m\Gamma x\|A^\rho\|\rho^{m-1}\Gamma'y\rangle = \sqrt{m[\Gamma]}\,\langle m\Gamma x|(m-1)\Gamma'y\rangle. \tag{16a}$$

Using eqs. (6a) and (9), we obtain from eq. (16a),

$$\langle\rho^{m-1}\Gamma'y\|B^\rho\|\rho^m\Gamma x\rangle = (-1)^{\Gamma'+\rho-\Gamma}\sqrt{m[\Gamma]}\,\langle m\Gamma x|(m-1)\Gamma'y\rangle. \tag{16b}$$

Note that the values of the reduced matrix elements of A^ρ and B^ρ differ only by a phase factor.

Using eq. (15) we can, in principle, generate an m-particle state from $|0\rangle$ by applying a product of m operators A. To produce a properly normalized state, it is more convenient to define a state creation operator $Z^\Gamma(mx)$ such that when it acts on the vacuum it produces a normalized and antisymmetrized right-hand state (or ket) of m particles,

$$Z^\Gamma(mx)|0\rangle = |\rho^m\Gamma x\rangle. \tag{17}$$

Its adjoint, $\overline{Z}^{\Gamma}(mx)$, produces a left-hand state or bra of m particles,

$$\langle \rho^m \Gamma \mu x| = \langle 0| \left(Z_\mu^\Gamma(mx)\right)^\dagger = (-1)^{\Gamma+\mu} \langle 0|\overline{Z}_{-\mu}^\Gamma(mx).$$

As examples, we have for $m = 1$,

$$\overline{Z}^\rho(1) = B^\rho \qquad \text{and} \qquad Z^\rho(1) = A^\rho,$$

and for $m = 2$,

$$\overline{Z}^\Gamma(2) = \frac{1}{\sqrt{2}} \left(B^\rho \times B^\rho\right)^\Gamma \qquad \text{and} \qquad Z^\Gamma(2) = \frac{-1}{\sqrt{2}} \left(A^\rho \times A^\rho\right)^\Gamma,$$

where $(-1)^\Gamma = -1$ to satisfy the antisymmetry requirement.

A matrix element can now be written in terms of the vacuum expectation value of the coupled product of second-quantized spherical-tensor operators. In the space of a single active orbit,

$$\langle \psi^\Gamma(mx) \| T^k \| \phi^{\Gamma'}(m'x') \rangle$$

$$= (-1)^{\Gamma+k-\Gamma'} \langle 0| \left(\overline{Z}^\Gamma(mx) \times \left(T^k \times Z^{\Gamma'}(m'x')\right)^\Gamma\right)^0 |0\rangle. \quad (18)$$

The angular momentum coupling between the different tensors is immaterial here since only the scalar part of the entire product can contribute to the vacuum expectation value and there is only one way to couple the three tensors to a scalar.

To understand the phase factor in eq. (18) it is best to go back to the single-bar matrix element and express it in terms of the vacuum expectation value,

$$\langle \psi_\mu^\Gamma(mx) | T_q^k | \phi_\nu^{\Gamma'}(m'x') \rangle = \langle 0|(Z_\mu^\Gamma)^\dagger(mx) T_q^k Z_\nu^{\Gamma'}(m'x') |0\rangle$$

$$= (-1)^{\Gamma+\mu} \langle 0|\overline{Z}_{-\mu}^\Gamma(mx) T_q^k Z_\nu^{\Gamma'}(m'x') |0\rangle$$

$$= (-1)^{\Gamma+\mu} \sum_{p\eta p'\eta'} \langle \Gamma, -\mu, p, \eta | p'\eta' \rangle \langle kq\Gamma'\nu|p\eta \rangle$$

$$\times \langle 0|(\overline{Z}^\Gamma \times (T^k \times Z^{\Gamma'})^p)_{\eta'}^{p'}|0\rangle.$$

Since only the $p' = 0$ part of the product can contribute, we have $p = \Gamma$ and $\eta = \mu$. Making use of eq. (B-2) and of

$$\langle \Gamma - \mu \Gamma \mu | 00 \rangle = (-1)^{\Gamma+\mu}[\Gamma]^{-\frac{1}{2}}, \quad (19)$$

we obtain

$$\langle \psi_\mu^\Gamma(mx) | T_q^k | \phi_\nu^{\Gamma'}(m'x') \rangle = (-1)^{\Gamma+\mu} \frac{(-1)^{\Gamma+\mu}}{[\Gamma]^{\frac{1}{2}}} (-1)^{k-\Gamma'+\mu}[\Gamma]^{\frac{1}{2}}$$

$$\times \begin{pmatrix} k & \Gamma' & \Gamma \\ q & \nu & -\mu \end{pmatrix} \langle 0|(\overline{Z}^\Gamma \times (T^k \times Z^{\Gamma'})^\Gamma)^0|0\rangle.$$

The phase factor of eq. (18) is obtained on comparing the result with eq. (3).

The reduced matrix element of a product, $(T^r \times U^s)^t$, of two tensors can be decomposed into a product of the reduced matrix elements of the two tensors,

$$\langle \psi^\Gamma(mx) \| (T^r \times U^s)^t \| \phi^{\Gamma'}(m'x') \rangle = (-1)^{\Gamma+\Gamma'+t} \sum_{m_0 \Gamma_0 y} [t]^{\frac{1}{2}} \begin{Bmatrix} \Gamma & r & \Gamma_0 \\ s & \Gamma' & t \end{Bmatrix}$$

$$\times \langle \psi^\Gamma(mx) \| T^r \| \xi^{\Gamma_0}(m_0 y) \rangle \langle \xi^{\Gamma_0}(m_0 y) \| U^s \| \phi^{\Gamma'}(m'x') \rangle. \quad (20)$$

To derive this result, we start with eq. (18) and use eq. (B-5) to recouple the tensors,

$$\langle \psi^\Gamma(mx) \| (T^r \times U^s)^t \| \phi^{\Gamma'}(m'x') \rangle$$

$$= (-1)^{\Gamma+t-\Gamma'} \langle 0| \left(\left(\overline{Z}^\Gamma(mx) \times (T^r \times U^s)^t \right)^{\Gamma'} \times Z^{\Gamma'}(m'x') \right)^0 |0 \rangle$$

$$= (-1)^{\Gamma+t-\Gamma'} \sum_{\Gamma_0} (-1)^{\Gamma+r+s+\Gamma'} [t\Gamma_0]^{\frac{1}{2}} \begin{Bmatrix} \Gamma & r & \Gamma_0 \\ s & \Gamma' & t \end{Bmatrix}$$

$$\times \langle 0| \left(\left((\overline{Z}^\Gamma(mx) \times T^r)^{\Gamma_0} \times U^s \right)^{\Gamma'} \times Z^{\Gamma'}(m'x') \right)^0 |0 \rangle$$

$$= (-1)^{r+s-t} (-1)^{2\Gamma'} \sum_{\Gamma_0} [t\Gamma_0]^{\frac{1}{2}} \begin{Bmatrix} \Gamma & r & \Gamma_0 \\ s & \Gamma' & t \end{Bmatrix}$$

$$\times \langle 0| \left((\overline{Z}^\Gamma(mx) \times T^r)^{\Gamma_0} \times (U^s \times Z^{\Gamma'}(m'x'))^{\Gamma_0} \right)^0 |0 \rangle.$$

Using the closure property,

$$\sum_{m_0 \Gamma_0'' \eta' y} |\chi_{\eta'}^{\Gamma_0''}(m_0 y) \rangle \langle \chi_{\eta'}^{\Gamma_0''}(m_0 y)| = 1,$$

we can introduce a complete set of states in the middle of the product of four tensors, and with the help of eq. (19) we obtain

$$\langle 0| \left((\overline{Z}^\Gamma(mx) \times T^r)^{\Gamma_0} \times (U^s \times Z^{\Gamma'}(m'x'))^{\Gamma_0} \right)^0 |0 \rangle$$

$$= \sum_\eta \frac{(-1)^{\Gamma_0-\eta}}{[\Gamma_0]^{\frac{1}{2}}} \sum_{m_0 \Gamma_0' \eta' y} \langle 0| (\overline{Z}^\Gamma(mx) \times T^r)_\eta^{\Gamma_0} |\chi_{\eta'}^{\Gamma_0'}(m_0 y) \rangle$$

$$\times \langle \chi_{\eta'}^{\Gamma_0'}(m_0 y)| (U^s \times Z^{\Gamma'}(m'x'))_{-\eta}^{\Gamma_0} |0 \rangle$$

$$= \sum_\eta \frac{(-1)^{\Gamma_0-\eta}}{[\Gamma_0]^{\frac{1}{2}}} \sum_{m_0 \Gamma_0' \eta' y} \langle 0| (\overline{Z}^\Gamma(mx) \times T^r)_\eta^{\Gamma_0} Z_{\eta'}^{\Gamma_0'}(m_0 y) |0 \rangle$$

$$\times (-1)^{\Gamma_0'+\eta'} \langle 0| \overline{Z}_{-\eta'}^{\Gamma_0'}(m_0 y) (U^s \times Z^{\Gamma'}(m'x'))_{-\eta}^{\Gamma_0} |0 \rangle.$$

Since only the scalar part can contribute to each of the two vacuum expectation values, we must have $\Gamma_0 = \Gamma_0'$ and $\eta = -\eta'$. On coupling the angular momenta of the tensors within each matrix element and making use of eq. (19) twice, we obtain

$$\langle 0| \left(\left(\overline{Z}^\Gamma(mx) \times T^r \right)^{\Gamma_0} \times \left(U^s \times Z^{\Gamma'}(m'x') \right)^{\Gamma_0} \right)^0 |0\rangle$$

$$= \sum_\eta \frac{(-1)^{\Gamma_0-\eta}}{[\Gamma_0]^{\frac{1}{2}}} \frac{(-1)^{\Gamma_0-\eta}}{[\Gamma_0]^{\frac{1}{2}}} \langle 0| \left(\overline{Z}^\Gamma(mx) \times T^r \times Z^{\Gamma_0}(m_0y) \right)^0 |0\rangle$$

$$\times (-1)^{\Gamma_0-\eta} \frac{(-1)^{\Gamma_0-\eta}}{[\Gamma_0]^{\frac{1}{2}}} \langle 0| \left(\overline{Z}^{\Gamma_0}(m_0y) \times U^s \times Z^{\Gamma'}(m'x') \right)^0 |0\rangle.$$

Since the final expression no longer contains η, the summation over η gives a factor $[\Gamma_0]$ and eq. (20) is obtained.

With two active orbits, a state $|((\rho_1^{m_1}\Gamma_1x_1) \times (\rho_2^{m_2}\Gamma_2x_2))^\Gamma\rangle$, formed of m_1 particles in orbit ρ_1 and m_2 particles in ρ_2, can be associated with a state creation operator $Z^\Gamma((\rho_1^{m_1}\Gamma_1x_1) \times (\rho_2^{m_2}\Gamma_2x_2))$ in the following way,

$$|((\rho_1^{m_1}\Gamma_1x_1) \times (\rho_2^{m_2}\Gamma_2x_2))^\Gamma\rangle = Z^\Gamma((\rho_1^{m_1}\Gamma_1x_1) \times (\rho_2^{m_2}\Gamma_2x_2))|0\rangle,$$

as in eq. (17). Since the state is a product of two parts, the operator can be decomposed similarly into a product of two operators, one involving only ρ_1 and the other only ρ_2. For later convenience, it is better to define the decomposition in terms of the adjoint state creation operator,

$$\overline{Z}^\Gamma\left((\rho_1^{m_1}\Gamma_1x_1) \times (\rho_2^{m_2}\Gamma_2x_2)\right) = \left(\overline{Z}^{\Gamma_1}(m_1x_1) \times \overline{Z}^{\Gamma_2}(m_2x_2)\right)^\Gamma. \qquad (21)$$

The decomposition for Z^Γ is obtained from the above expression in an analogous way as for eq. (7),

$$Z^\Gamma\left((\rho_1^{m_1}\Gamma_1x_1) \times (\rho_2^{m_2}\Gamma_2x_2)\right) = (-1)^{\Gamma_1+\Gamma_2+\Gamma}\left(Z^{\Gamma_2}(m_2x_2) \times Z^{\Gamma_1}(m_1x_1)\right)^\Gamma$$

$$= (-1)^{m_1m_2}\left(Z^{\Gamma_1}(m_1x_1) \times Z^{\Gamma_2}(m_2x_2)\right)^\Gamma, \qquad (22)$$

where the dependence on m_1m_2 in the phase factor arises from the need of commuting m_1 creation operators in Z^{Γ_1} to the left of Z^{Γ_2}.

With eqs. (21) and (22), we can now derive the general relation for the reduced matrix element of a product of two tensor operators each operating in a different orbit. Let the two orbits be distinguished by numerals 1 and 2. We can resolve the reduced matrix element of the product tensor into those of its constituents:

$$\langle(\phi^{\Gamma_1}(m_1x_1) \times \phi^{\Gamma_2}(m_2x_2))^{\Gamma}\|(T^r(1) \times U^s(2))^t\|(\psi^{\Gamma_1'}(m_1'x_1') \times \psi^{\Gamma_2'}(m_2'x_2'))^{\Gamma'}\rangle$$

$$= (-1)^{\Gamma-\Gamma'+t}(-1)^{m_1'm_2'}\langle 0|\left((\overline{Z}^{\Gamma_1}(m_1x_1) \times \overline{Z}^{\Gamma_2}(m_2x_2))^{\Gamma}\right.$$
$$\times (T^r(1) \times U^s(2))^t \times (Z^{\Gamma_1'}(m_1'x_1') \times Z^{\Gamma_2'}(m_2'x_2'))^{\Gamma'}\Big)^0|0\rangle$$

$$= (-1)^{\Gamma-\Gamma'+t}(-1)^{m_1'm_2'}\sum_{\gamma_1\gamma_2}[\Gamma t\gamma_1\gamma_2]^{\frac{1}{2}}\begin{Bmatrix}\Gamma_1 & \Gamma_2 & \Gamma \\ r & s & t \\ \gamma_1 & \gamma_2 & \Gamma'\end{Bmatrix}$$
$$\times (-1)^{m_2(m_1-m_1')}\langle 0|(\overline{Z}^{\Gamma_1}(m_1x_1) \times T^r(1))^{\gamma_1}$$
$$\times (\overline{Z}^{\Gamma_2}(m_2x_2) \times U^s(2))^{\gamma_2} \times (Z^{\Gamma_1'}(m_1'x_1') \times Z^{\Gamma_2'}(m_2'x_2'))^{\Gamma'}|0\rangle$$

$$= (-1)^{\Gamma-\Gamma'+t}(-1)^{m_1'm_2'}\sum_{\gamma_1\gamma_2}[\Gamma t\gamma_1\gamma_2]^{\frac{1}{2}}\begin{Bmatrix}\Gamma_1 & \Gamma_2 & \Gamma \\ r & s & t \\ \gamma_1 & \gamma_2 & \Gamma'\end{Bmatrix}$$
$$\times (-1)^{m_2(m_1-m_1')}\sum_{\gamma'}[\Gamma'\gamma']\begin{Bmatrix}\gamma_1 & \gamma_2 & \Gamma' \\ \Gamma_1' & \Gamma_2' & \Gamma' \\ \gamma' & \gamma' & 0\end{Bmatrix}(-1)^{m_1'm_2'}$$
$$\times \langle 0|\left((\overline{Z}^{\Gamma_1}(m_1x_1) \times T^r(1))^{\gamma_1} \times Z^{\Gamma_1'}(m_1'x_1')\right)^{\gamma'}$$
$$\times \left((\overline{Z}^{\Gamma_2}(m_2x_2) \times U^s(2))^{\gamma_2} \times Z^{\Gamma_2'}(m_2'x_2')\right)^{\gamma'}|0\rangle.$$

Since within each orbit only the scalar part can contribute to the vacuum expectation value, we have $\gamma' = 0$. As a result, $\gamma_1 = \Gamma_1'$, $\gamma_2 = \Gamma_2'$, and the second 9j-symbol reduces to

$$\begin{Bmatrix}\gamma_1 & \gamma_2 & \Gamma' \\ \Gamma_1' & \Gamma_2' & \Gamma' \\ \gamma' & \gamma' & 0\end{Bmatrix} = \begin{Bmatrix}\Gamma_1' & \Gamma_2' & \Gamma' \\ \Gamma_1' & \Gamma_2' & \Gamma' \\ 0 & 0 & 0\end{Bmatrix} - \frac{1}{[\Gamma_1'\Gamma_2'\Gamma']^{\frac{1}{2}}},$$

using eqs. (B-11) and (B-21). The final result is

$$\langle(\phi^{\Gamma_1}(m_1x_1) \times \phi^{\Gamma_2}(m_2x_2))^{\Gamma}\|(T^r(1) \times U^s(2))^t\|(\psi^{\Gamma_1'}(m_1'x_1') \times \psi^{\Gamma_2'}(m_2'x_2'))^{\Gamma'}\rangle$$

$$= (-1)^{\Gamma-\Gamma'+t}(-1)^{m_2(m_1-m_1')}[\Gamma\Gamma't]^{\frac{1}{2}}\begin{Bmatrix}\Gamma_1 & \Gamma_2 & \Gamma \\ r & s & t \\ \Gamma_1' & \Gamma_2' & \Gamma'\end{Bmatrix}$$
$$\times \langle 0|\overline{Z}^{\Gamma_1}(m_1x_1) \times T^r(1) \times Z^{\Gamma_1'}(m_1'x_1')|0\rangle$$
$$\times \langle 0|\overline{Z}^{\Gamma_2}(m_2x_2) \times U^s(2) \times Z^{\Gamma_2'}(m_2'x_2')|0\rangle$$

$$= (-1)^{m_2(m_1-m_1')}[\Gamma\Gamma't]^{\frac{1}{2}}\begin{Bmatrix}\Gamma_1 & \Gamma_2 & \Gamma \\ \Gamma_1' & \Gamma_2' & \Gamma' \\ r & s & t\end{Bmatrix}$$
$$\times \langle\phi^{\Gamma_1}(m_1x_1)\|T^r(1)\|\psi^{\Gamma_1'}(m_1'x_1')\rangle\langle\phi^{\Gamma_2}(m_2x_2)\|U^s(2)\|\psi^{\Gamma_2'}(m_2'x_2')\rangle, \quad (23)$$

where we have interchanged two rows of the $9j$-symbol to absorb the angular momentum phase factors. The particle-number dependent phase factor is different from $+1$ only when T^r and U^s are not number conserving operators.

Using eq. (23) we can derive several other tensorial identities. If $t = 0$ (and $r = s$), the $9j$-symbol reduces to a $6j$-symbol by eq. (B-20), and we obtain

$$\langle (\phi^{\Gamma_1}(m_1 x_1) \times \phi^{\Gamma_2}(m_2 x_2))^{\Gamma} \| (T^r(1) \times U^r(2))^0 \| (\psi^{\Gamma_1'}(m_1' x_1') \times \psi^{\Gamma_2'}(m_2' x_2'))^{\Gamma} \rangle$$

$$= (-1)^{m_2(m_1 - m_1')}(-1)^{\Gamma_2 + \Gamma_1' + \Gamma - r} \left[\frac{\Gamma}{r}\right]^{\frac{1}{2}} \left\{ \begin{matrix} \Gamma_1 & \Gamma_2 & \Gamma \\ \Gamma_2' & \Gamma_1' & r \end{matrix} \right\}$$

$$\times \langle \phi^{\Gamma_1}(m_1 x_1) \| T^r(1) \| \psi^{\Gamma_1'}(m_1' x_1') \rangle \langle \phi^{\Gamma_2}(m_2 x_2) \| U^r(2) \| \psi^{\Gamma_2'}(m_2' x_2') \rangle. \qquad (24)$$

If $U^s = 1$, eq. (23) reduces to

$$\langle (\phi^{\Gamma_1}(m_1 x_1) \times \phi^{\Gamma_2}(m_2 x_2))^{\Gamma} \| T^r(1) \| (\psi^{\Gamma_1'}(m_1' x_1') \times \phi^{\Gamma_2}(m_2 x_2))^{\Gamma'} \rangle$$

$$= (-1)^{m_2(m_1 - m_1')}(-1)^{\Gamma_1 + \Gamma_2 + \Gamma' + r}[\Gamma \Gamma']^{\frac{1}{2}} \left\{ \begin{matrix} \Gamma_1 & \Gamma & \Gamma_2 \\ \Gamma' & \Gamma_1' & r \end{matrix} \right\}$$

$$\times \langle \phi^{\Gamma_1}(m_1 x_1) \| T^r(1) \| \psi^{\Gamma_1'}(m_1' x_1') \rangle, \qquad (25)$$

where we have used the fact that $\langle \Gamma \| 1 \| \Gamma \rangle = [\Gamma]^{\frac{1}{2}}$. On the other hand, if $T^r = 1$, eq. (23) becomes

$$\langle (\phi^{\Gamma_1}(m_1 x_1) \times \phi^{\Gamma_2}(m_2 x_2))^{\Gamma} \| U^s(2) \| (\phi^{\Gamma_1}(m_1 x_1) \times \psi^{\Gamma_2'}(m_2' x_2'))^{\Gamma'} \rangle$$

$$= (-1)^{\Gamma_1 + \Gamma_2' + \Gamma + s}[\Gamma \Gamma']^{\frac{1}{2}} \left\{ \begin{matrix} \Gamma_2 & \Gamma & \Gamma_1 \\ \Gamma' & \Gamma_2' & s \end{matrix} \right\} \langle \phi^{\Gamma_2}(m_2 x_2) \| U^s(2) \| \psi^{\Gamma_2'}(m_2' x_2') \rangle.$$

$$(26)$$

Eqs. (25) and (26) are useful to remove the part of a state that is not acted upon by the operator in a reduced matrix element involving several orbits.

Appendix D

Commonly Used Elementary Operators

We describe here some of the commonly used elementary operators and express them in terms of the angular momentum coupled products of single-particle creation and annihilation operators. Where possible, the values of the defining matrix elements and unitary decompositions of the operators are also given.

D.1 Particle transfer operators

The operators for adding and taking away a particle in orbit r are A^r and B^r, respectively. Their reduced matrix elements are related to the coefficient of fractional parentage through eqs. (C-15) and (C-16). In terms of spectroscopic factors,

$$S_{\text{stripping}} = \frac{1}{[\Gamma]} |\langle (m+1)\Gamma x \| A^r \| m\Gamma_0 y \rangle|^2, \qquad (1)$$

for stripping, and

$$S_{\text{pickup}} = \frac{1}{[\Gamma]} |\langle (m-1)\Gamma_0 y \| B^r \| m\Gamma x \rangle|^2, \qquad (2)$$

for pickup reactions. The operators have particle rank $k = \frac{1}{2}$, and unitary rank $(\mu, \nu) = (\frac{1}{2}, \frac{1}{2})$ for A^r and $(\frac{1}{2}, -\frac{1}{2})$ for B^r.

For two-nucleon transfer reactions, the amplitude is given by the matrix elements between target and final states of operator $(A^r \times A^r)^\Delta$ for the stripping and $(B^r \times B^r)^\Delta$ for the pickup of a pair of particles in orbit r. The particle rank of these operators is $k = 1$, and the unitary rank $(\mu, \nu) = (1, 1)$ for stripping and $(1, -1)$ for pickup.

The more complicated nucleon-cluster transfer operators consist of products of several single-particle creation or annihilation operators. The angular momenta are coupled appropriately to reflect the structure of the cluster transferred. For example, an α-particle transfer operator can be written as $((A^r \times A^r)^\Delta \times (A^r \times A^r)^\Delta)^0$ with the intermediate spin-isospin $(\Delta_J, \Delta_T) = (0, 1)$ corresponding to the fact that both the proton and the neutron pairs are zero-coupled in an α-particle.

D.2   Number operator

The number operator for particles in orbit r can be written in the form

$$\hat{n}_r = [r]^{\frac{1}{2}}(A^r \times B^r)^0. \tag{3}$$

Although the particle rank of this operator is $k = 1$, the unitary rank is $(0,0)$ in configuration space.

One of the main uses of the number operator is in the one-body part of the Hamiltonian,

$$H(1) = \sum_r \epsilon_r \hat{n}_r, \tag{4}$$

where ϵ_r is the single-particle energy. In scalar space, it can be decomposed into two parts as described in Chapter V,

$$H(1) = \bar{\epsilon}\hat{n} + \sum_r (\epsilon_r - \bar{\epsilon})\hat{n}_r, \tag{5}$$

where the average single-particle energy

$$\bar{\epsilon} = \frac{1}{N} \sum_r N_r \epsilon_r,$$

was given in eq. (IV-7) as the defining quantity for the unitary rank $(0,0)$ part of $H(1)$, and the traceless single-particle energy

$$\tilde{\epsilon}_r = (\epsilon_r - \bar{\epsilon}), \tag{6}$$

is the defining matrix element of the unitary rank $(1,0)$ part as given by eq. (V-23).

D.3   One-body electromagnetic transition operators

The electric λ-th multipole transition operator in the long wavelength approximation is a one-body operator and can be written in the form

$$\hat{O}^{E\lambda} = \sum_{rs} v_{rs}^{E\lambda} \frac{(A^r \times B^s)^\lambda}{[\lambda]^{\frac{1}{2}}}, \tag{7}$$

where the single-particle value $v_{rs}^{E\lambda}$ is given by

$$v_{rs}^{E\lambda} = \langle r\|\hat{O}^{E\lambda}\|s\rangle = \langle r\|r^\lambda Y^\lambda(\theta,\phi)(\tfrac{1}{2} + \tau_z)\|s\rangle. \tag{8}$$

The factor $(\tfrac{1}{2} + \tau_z)$ projects out protons. The operator is a mixture of isoscalar and isovector parts. Except for $\lambda = 0$, the particle rank is $k = 1$ and the unitary ranks are $(\mu,\nu) = (1,0)$ in scalar averaging.

The isoscalar, single-particle value is given by

$$v_{rs}^{E\lambda,0} = \frac{1}{\sqrt{2}} \langle r \| r^\lambda Y^\lambda \| s \rangle, \tag{9}$$

and the isovector part by

$$v_{rs}^{E\lambda,1} = \sqrt{\frac{3}{2}} \langle r \| r^\lambda Y^\lambda \| s \rangle. \tag{10}$$

In eqs. (9) and (10) the reduced matrix elements do not involve isospin.

The reduced matrix element $\langle r \| r^\lambda Y^\lambda \| s \rangle$ between orbits r and s can be separated into a product of a radial and an angular part,

$$\langle r \| r^\lambda Y^\lambda \| s \rangle = \langle r^\lambda \rangle_{rs} \langle j_r \| Y^\lambda \| j_s \rangle, \tag{11}$$

where the radial integral is given by

$$\langle r^\lambda \rangle_{rs} = \int R_{n_r l_r}^*(r) \, r^\lambda R_{n_s l_s}(r) r^2 \, dr.$$

Using eq. (C-25) and the coupling order $j = l + s$ (vectorially),

$$\langle j_r \| Y^\lambda \| j_s \rangle = (-1)^{j_s + \lambda + \frac{1}{2}} \sqrt{\frac{[j_r j_s l_r l_s \lambda]}{4\pi}} \begin{pmatrix} l_r & \lambda & l_s \\ 0 & 0 & 0 \end{pmatrix} \begin{Bmatrix} l_r & j_r & \frac{1}{2} \\ j_s & l_s & \lambda \end{Bmatrix}. \tag{12}$$

An analogous expression is

$$\langle j_r \| Y^\lambda \| j_s \rangle = (-1)^{j_r + \frac{1}{2}} \sqrt{\frac{[j_r \lambda j_s]}{4\pi}} \begin{pmatrix} j_r & \lambda & j_s \\ \frac{1}{2} & 0 & -\frac{1}{2} \end{pmatrix} \frac{1}{2} \left[1 + (-1)^{l_r + \lambda + l_s} \right], \tag{12a}$$

which does not involve any $6j$-symbol.

Instead of using the isospin formalism, one can work in terms of protons and neutrons. The single-particle value becomes

$$v_{rs}^{E\lambda} = \langle j_r \| r^\lambda Y^\lambda \| j_s \rangle, \tag{13}$$

if both r and s are proton orbitals, and zero otherwise.

The single-particle value for magnetic λ-th multipole is given by

$$v_{rs}^{M\lambda} = \langle r \| \hat{O}^{M\lambda} \| s \rangle = \mu_N \langle r \| \nabla \left(r^\lambda Y_M^\lambda(\theta, \phi) \right) \cdot \left[\frac{1}{2} g^s \mathbf{s} + g^l \left(\frac{\mathbf{l}}{\lambda + 1} \right) \right] \| s \rangle, \tag{14}$$

where $\mu_N = \hbar e / 2 M_p$ is the nuclear magneton and the gyromagnetic ratios are given by

$$g^l = \begin{cases} g_p^l = 1 & \text{for protons,} \\ g_n^l = 0 & \text{for neutrons,} \end{cases} \quad \text{and} \quad g^s = \begin{cases} g_p^s = 5.5857 & \text{for protons,} \\ g_n^s = -3.8263 & \text{for neutrons.} \end{cases}$$

The single-particle value in units of μ_N is

$$v_{rs}^{M\lambda,T} = (-1)^T \sqrt{\frac{[T]}{2}} \langle r^{\lambda-1} \rangle_{rs} (-1)^{l_r} [\lambda] \sqrt{\frac{\lambda(2\lambda-1)[l_r l_s j_r j_s]}{4\pi}} \begin{pmatrix} l_r & \lambda-1 & l_s \\ 0 & 0 & 0 \end{pmatrix}$$

$$\times \left[2\frac{g_p^l + (-1)^T g_n^l}{\lambda+1}(-1)^{l_s+j_s+\frac{1}{2}} \sqrt{l_s(l_s+1)[l_s]} \begin{Bmatrix} l_r & l_s & \lambda \\ j_s & j_r & \frac{1}{2} \end{Bmatrix} \begin{Bmatrix} \lambda-1 & 1 & \lambda \\ l_s & l_r & l_s \end{Bmatrix} \right.$$

$$\left. + \sqrt{\frac{3}{2}}\{g_p^s + (-1)^T g_n^s\} \begin{Bmatrix} l_r & \frac{1}{2} & j_r \\ l_s & \frac{1}{2} & j_s \\ \lambda-1 & 1 & \lambda \end{Bmatrix} \right], \tag{15}$$

where $T = 0$ for isoscalar and $T = 1$ for isovector parts.

In terms of neutron and proton space,

$$v_{rs}^{M\lambda} = \langle r^{\lambda-1} \rangle_{rs}(-1)^{l_r}[\lambda]\sqrt{\frac{\lambda(2\lambda-1)[l_r l_s j_r j_s]}{4\pi}} \begin{pmatrix} l_r & \lambda-1 & l_s \\ 0 & 0 & 0 \end{pmatrix}$$

$$\times \left[\frac{2}{\lambda+1}(-1)^{l_s+j_s+\frac{1}{2}} \sqrt{l_s(l_s+1)[l_s]} \begin{Bmatrix} l_r & l_s & \lambda \\ j_s & j_r & \frac{1}{2} \end{Bmatrix} \begin{Bmatrix} \lambda-1 & 1 & \lambda \\ l_s & l_r & l_s \end{Bmatrix} \right.$$

$$\left. + g_p^s \sqrt{\frac{3}{2}} \begin{Bmatrix} l_r & \frac{1}{2} & j_r \\ l_s & \frac{1}{2} & j_s \\ \lambda-1 & 1 & \lambda \end{Bmatrix} \right], \tag{16}$$

for transitions between proton orbits, and

$$v_{rs}^{M\lambda} = \langle r^{\lambda-1} \rangle_{rs}(-1)^{l_r}[\lambda]\sqrt{\frac{\lambda(2\lambda-1)[l_r l_s j_r j_s]}{4\pi}}$$

$$\times \begin{pmatrix} l_r & \lambda-1 & l_s \\ 0 & 0 & 0 \end{pmatrix} g_n^s \sqrt{\frac{3}{2}} \begin{Bmatrix} l_r & \frac{1}{2} & j_r \\ l_s & \frac{1}{2} & j_s \\ \lambda-1 & 1 & \lambda \end{Bmatrix}, \tag{17}$$

for transitions between neutron orbits. The single particle values between proton and neutron orbits vanish.

D.4 Gamow-Teller operator for allowed beta-decay

Beta-decay transforms a proton into a neutron and vice versa. The Fermi part is simply the single-particle isospin raising or lowering operator. The Gamow-Teller part involves the spin and can be written in neutron-proton space as

$$\hat{O}^{GT} = \hat{\sigma}. \tag{18}$$

It is a $k = 1$, $(\mu, \nu) = (1, 0)$ operator with single-particle value $\sqrt{6}$ if it operates between a proton and a neutron orbit, and zero otherwise. In isospin space, the operator can be written as

$$\hat{O}^{GT} = \hat{\sigma}\hat{\tau}, \tag{19}$$

and the single-particle value is

$$v_{rs}^{GT} = (-1)^{l_r + j_r + \frac{3}{2}} 6 [j_r j_s]^{\frac{1}{2}} \begin{Bmatrix} j_r & \frac{1}{2} & l_r \\ \frac{1}{2} & j_s & 1 \end{Bmatrix} \delta_{l_r l_s}, \tag{20}$$

where l_r and l_s are the orbital angular momenta of orbits r and s respectively.

D.5   Scalar two-body operator

The most commonly encountered angular momentum scalar two-body operator is the two-body part of a Hamiltonian. It can be written in the form

$$\hat{O}(2) = - \sum_{rstu\Gamma} \varsigma_{rs}\varsigma_{tu}[\Gamma]^{\frac{1}{2}} W_{rstu}^{\Gamma} \left((A^r \times A^s)^{\Gamma} \times (B^t \times B^u)^{\Gamma} \right)^{00}, \tag{21}$$

where $\varsigma_{rr'} = 1/\sqrt{1 + \delta_{rr'}}$, and $W_{rstu}^{\Gamma} \equiv \langle rs\Gamma|V|tu\Gamma\rangle$ is the normalized and antisymmetrized two-body matrix element with the following symmetries:

$$W_{rstu}^{JT} = -(-1)^{r+s-J-T} W_{srtu}^{JT}$$
$$= -(-1)^{t+u-J-T} W_{rsut}^{JT} = (-1)^{r+s-t-u} W_{srut}^{JT}. \tag{22}$$

Alternatively, we can write

$$\hat{O}(2) = \sum_{rstu\Delta} \varsigma_{rs}\varsigma_{tu}\beta_{rtsu}^{\Delta} \left((A^r \times B^t)^{\Delta} \times (A^s \times B^u)^{\Delta} \right)^{00}$$
$$+ \sum_{rst} \varsigma_{rs}\varsigma_{st}[s]^{\frac{1}{2}} \beta_{rtss}^{0} (A^r \times B^t)^{0}, \tag{23}$$

where the multipole coefficient,

$$\beta_{rtsu}^{\Delta} = \sum_{\Gamma} (-1)^{s+t+\Delta+\Gamma} [\Delta]^{\frac{1}{2}} [\Gamma] \begin{Bmatrix} r & s & \Gamma \\ u & t & \Delta \end{Bmatrix} W_{rstu}^{JT}, \tag{24}$$

having the symmetry property

$$\beta_{rtsu}^{\Delta} = \beta_{surt}^{\Delta} = (-1)^{r+u-s-t} \beta_{trus}^{\Delta},$$

is another way to write the defining matrix elements of an angular momentum scalar two-body operator.

The operator $\hat{O}(2)$ has particle rank $k = 2$ and mixed unitary rank up to $(\mu, \nu) = (2, 0)$. For scalar averaging, the number operator is written as

$$\hat{n} = \sum_r [r]^{\frac{1}{2}} (A^r \times B^r)^0,$$

and the unitary rank $(0,0)$ part of $\hat{O}(2,0)$ can be shown to be

$$\hat{O}_2(0,0) = \overline{W} \binom{\hat{n}}{2}, \qquad \text{with} \qquad \overline{W} = \binom{N}{2}^{-1} \sum_{r \leq s} \sum_{\Gamma} [\Gamma] W^{\Gamma}_{rsrs}. \qquad (25)$$

To extract the unitary rank $(1,0)$ part, we must first subtract the $O_2(0,0)$ part from $\hat{O}(2)$ and then apply all possible left-contractions to the remainder. The result for the $(\mu, \nu) = (1, 0)$ part is

$$\hat{O}_2(1,0) = \sum_{rs} \overline{W}_{rs} (\hat{n} - 1) [r]^{\frac{1}{2}} (A^r \times B^s)^0, \qquad (26)$$

with

$$\overline{W}_{rs} = \frac{1}{N_r(N-2)} \sum_t \sum_{\Gamma} [\Gamma] W^{\Gamma}_{rtst} \varsigma^{-1}_{rt} \varsigma^{-1}_{st} - \overline{W} \delta_{rs}.$$

The $(\mu, \nu) = (2, 0)$ part is given by

$$\hat{O}_2(2,0) = - \sum_{\substack{r \leq s \\ t \leq u \\ \Gamma}} \varsigma_{rs} \varsigma_{tu} [\Gamma]^{\frac{1}{2}} \mathcal{W}^{\Gamma}_{rstu} \left((A^r \times A^s)^{\Gamma} \times (B^t \times B^u)^{\Gamma} \right)^{00}, \qquad (27)$$

with

$$\begin{aligned}
\mathcal{W}^{\Gamma}_{rsrs} &= W^{\Gamma}_{rsrs} - \overline{W}_{rr} - \overline{W}_{ss} - \overline{W} && \text{for diagonal elements,} \\
\mathcal{W}^{\Gamma}_{rtst} &= W^{\Gamma}_{rtst} - \overline{W}_{rs} \varsigma^{-1}_{rt} \varsigma^{-1}_{st} && \text{for } r \neq s \text{ but } j^{\pi}_r = j^{\pi}_s, \\
\mathcal{W}^{\Gamma}_{rstu} &= W^{\Gamma}_{rstu} && \text{otherwise,}
\end{aligned}$$

i.e., it is the remainder of the operator $\hat{O}(2)$ in eq. (21) after removing the contributions already included in eqs. (25) and (26).

The situation is slightly more complicated for configuration averaging since we must decompose the operator with respect to the unitary rank of each orbit. The unitary rank $(0,0)$ part is given by

$$\hat{O}_2(0,0) = \sum_{r \leq s} \overline{W}(r,s) \frac{\hat{n}_r (\hat{n}_s - \delta_{rs})}{1 + \delta_{rs}},$$

with

$$\overline{W}(r,s) = \frac{1+\delta_{rs}}{N_r(N_s-\delta_{rs})} \sum_{\Gamma} [\Gamma] W^{\Gamma}_{rsrs}, \tag{28}$$

and the $(1,0)$ part by

$$\hat{O}_2(1,0) = \sum_{rst} \overline{W}(r,s;t)\,\hat{n}_t\,[r]^{\frac{1}{2}}(A^r \times B^s)^0, \tag{29}$$

with

$$\overline{W}(r,s;t) = \frac{\sqrt{1+\delta_{rt}+\delta_{st}}}{N_r(N_t-\delta_{rt}-\delta_{st})} \sum_{\Gamma} [\Gamma] W^{\Gamma}_{rtst}, \tag{30}$$

for $r \neq s$ but $j_r^{\pi} = j_s^{\pi}$. Analogous to eq. (27), the two-body matrix elements for the configuration unitary rank $(2,0)$ part become

$$V^{\Gamma}_{rstu} = W^{\Gamma}_{rstu} - \frac{1+\delta_{rs}}{N_r(N_s-\delta_{rs})}\overline{\Omega}, \tag{31}$$

where

$$\overline{\Omega} = \begin{cases} \sum_{\Gamma}[\Gamma]W^{\Gamma}_{rstu} & \text{for } r=t,\ s=u \\ & \text{or } r=u,\ s=t; \\[2mm] \sum_{\Gamma}[\Gamma]W^{\Gamma}_{rstu} & \text{for } r=t,\ s\neq u \text{ but } j_r^{\pi}=j_s^{\pi}, \\ & \text{or } s=u,\ r\neq t \text{ but } j_s^{\pi}=j_u^{\pi}; \\[2mm] -\sum_{\Gamma}[\Gamma](-1)^{r+s-\Gamma}W^{\Gamma}_{rstu} & \text{for } r=u,\ s\neq t \text{ but } j_s^{\pi}=j_t^{\pi}, \\ & \text{or } s=t,\ r\neq u \text{ but } j_r^{\pi}=j_u^{\pi}. \end{cases} \tag{32}$$

Appendix E

Algorithm for Basic Diagram Generation

In Chapter V, it was mentioned that the identification of a complete set of basic diagrams is an important step in applying diagrammatic methods to trace calculations. By basic diagrams we mean a group of diagrams that cannot be transformed into each other by applying a given set of symmetry operations. Two criteria must therefore be satisfied by a set of basic diagrams:

    1) All the diagrams in the set are basic, and

    2) The set is complete.

The first criterion can be satisfied by applying all the symmetry operations to the product operator for each of the diagrams in the set and checking that none of them can be transformed into another one in the set. The second criterion is much harder to achieve since one must ensure that *all* the diagrams required in the trace calculation can be produced from the set.

The algorithm described here was given in Chang and Wong (1978), the first of a three part series in using a computer (i) to generate the basic diagrams (Chang and Wong, 1979), (ii) to derive the equations for evaluating the contribution of a diagram to the trace (Chang and Wong, 1980), and (iii) to write the Fortran code for the numerical calculation itself (Chang, Draayer and Wong, 1982). The primary aim of the discussion here is to illustrate the principles of basic diagram generation and the simplicity that can be achieved with a computer.

Each elementary operator $\hat{O}_i$ is taken to be made up of single-particle creation and annihilation operators angular momentum coupled in the following *fan-shaped pattern*

$$\hat{O}_i = \left(\cdots \left((\alpha_{1i} \times \alpha_{2i})^{\gamma_{2i}} \times \alpha_{3i} \right)^{\gamma_{3i}} \times \cdots \right)^{\Gamma_i}, \tag{1}$$

where α_{ji} can be either a single-particle creation operator A^r or an annihilation operator B^r. The elementary operators are in turn coupled together in a similar fan-shaped pattern to produce the product operator $\hat{O}$,

$$\hat{O} = \left(\cdots \left((\hat{O}_1^{\Gamma_1} \times \hat{O}_2^{\Gamma_2})^{\Delta_2} \times \hat{O}_3^{\Gamma_3} \right)^{\Delta_3} \times \cdots \right)^0. \tag{2}$$

The final angular momentum rank is taken to be zero since this is the only part that can contribute to scalar and configuration traces. None of the angular momentum arguments in $\hat{O}$ is needed in the search for basic diagrams; however, they constitute an essential part of the operator and are needed in evaluating the contribution of the diagram to the trace.

For diagram generation, we can represent the operator $\hat{O}$, with maximum particle rank p, as a product of $2p$ single-particle operators of the form

$$\hat{O} \rightarrow \alpha_1 \alpha_2 \cdots \alpha_{2p}, \tag{3}$$

with the first $2\nu_1$ factors coming from the elementary operator $\hat{O}_1$, the second $2\nu_2$ from $\hat{O}_2$, and so on. Each single-particle operator is identified by a sign and two indices. The sign is used to indicate whether it is a creation operator $(+)$ or an annihilation operator $(-)$. One of the two indices gives the position among the $2p$ single-particle operators in eq. (3), and the other indicates the elementary operator it belongs to.

A diagram is a particular set of p contractions among the $2p$ single-particle operators. We can represent such a diagram in a computer program by a set of p pairs of indices $\{l_i, r_i\}$, $i = 1, \ldots, p$, where l_i and r_i are the position indices of a pair of contracting single-particle operators. Let the pairs be arranged such that $l_i < r_i$, i.e., the left operator is always earlier than the right operator, as far as their position in eq. (3) is concerned. In this way we can recognize that the pair i is a left-contraction if α_{l_i} is a creation operator and is a right-contraction otherwise.

Once the p contractions are fixed, the order in which they are carried out is immaterial. We can therefore adopt a special ordering scheme among the p pairs of indices that is convenient for identifying basic diagrams. The one we shall use is $l_i < l_{i+1}$, i.e., the left indices are arranged in ascending order for all the contracting pairs in a diagram. This completes the set of rules necessary to represent a diagram unambiguously on a computer.

Next we shall define a relative order between different diagrams in such a way that a sorting procedure can be used to order a large number of diagrams. Given two diagrams, $\{l_i, r_i\}$ and $\{l_i', r_i'\}$, the unprimed one is the earlier one if $l_1 < l_1'$, and we must have $r_1 < r_1'$ if $l_1 = l_1'$. If, on the other hand, $l_i = l_i'$ and $r_i = r_i'$ for $i = 1$ to k, we must have $l_{k+1} < l_{k+1}'$, or if $l_{k+1} = l_{k+1}'$, then $r_{k+1} < r_{k+1}'$. In short, the sequence of p pairs of indices $l_1 r_1 l_2 r_2 \cdots l_p r_p$ is viewed as a multi-digit number, and the order between diagrams is simply the ascending (or descending if one so wishes) order between these multi-digit numbers representing the diagrams. In this way, we can easily make use of sorting techniques to order a large number of diagrams represented in terms of $\{l_i, r_i\}$, $i = 1, \ldots, p$.

It is easy to recognize a basic diagram in this scheme. Given a diagram represented by $\{l_i, r_i\}$, $i = 1, \ldots, p$, it is a basic diagram if all the diagrams derived from it by symmetry transformations are later than it according to the order defined in the previous paragraph. In general, one can check whether a diagram is a basic one by applying all the possible symmetry transformations to the product operator and seeing if any of the diagrams generated corresponds to one of the basic diagrams already found. The algorithm described here accomplishes the purpose without having to compare explicitly each diagram generated with a set of known basic diagrams. While the method may be too tedious for carrying out by hand, it is ideal for a computer.

Appendix F

Propagation of Fixed-JT Averages

We shall describe here a method to propagate fixed-JT averages using an iterative procedure. Since the process is more time-consuming than the one-step approach used in most other averages, we must devise a way to reduce the amount of computation that has to be carried out each time the fixed-JT averaging method is applied. This calls for the identification of those parts of calculations that can be isolated and are repeated often. The single-orbit traces appearing on the right-hand side of eq. (VIII-24) are such quantities, and we shall describe here a method to calculate them so that the results can be saved for subsequent usage.

The first step is to identify the basic components of all the operators that may enter in applications involving fixed-JT averaging. We shall assume that the most complicated physical operator that can be anticipated is the square of a Hamiltonian. Thus the highest particle rank operator we need is $k = 4$. By restricting ourselves to elementary operators that are angular momentum scalars, we need to consider only those $k = 4$ operators that are coupled to $(J, T) = (0, 0)$. This greatly reduces the number of averages that need to be stored. We shall consider also each j-orbit as a separate space and evaluate the single-orbit traces up to angular momentum scalar operators of particle rank $k = 4$.

Besides $k = 4$, we need also all the operators of lower particle ranks. These are no longer necessarily angular momentum scalars since they may constitute only a part of the product operators and they couple with other parts of the product operator in a different orbit to form an overall scalar operator. In summary, the following list of single-orbit operators are needed to evaluate the averages of product operators up to the complexity of H^2:

$$(A \times B)^t, \quad \left(((A \times A)^{\Delta_1} \times B)^{\Delta_2} \times B \right)^t,$$

$$\left((A \times (A \times A)^{\Delta_1})^{\Delta_2} \times (B \times (B \times B))^{\Delta_3} \right)^t,$$

$$\left(((A \times A)^{\Delta_1} \times (B \times B)^{\Delta_1})^0 \times ((A \times A)^{\Delta_2} \times (B \times B)^{\Delta_2})^0 \right)^0.$$

As usual with calculations involving both spin and isospin, we shall use a single Greek letter to stand for both spin and isospin throughout this appendix.

Let us start by deriving the propagation equation. For a diagonal matrix element of a product of two spherical tensors, eq. (C-20) can be written in the form

$$\langle m\Gamma x\|(P^r \times Q^s)^t\|m\Gamma x\rangle = \sum_{\Gamma_0 y}(-1)^{2\Gamma+t}[t]^{\frac{1}{2}}\begin{Bmatrix}\Gamma & r & \Gamma_0 \\ s & \Gamma & t\end{Bmatrix}$$

$$\times \langle m\Gamma x\|P^r\|m_0\Gamma_0 y\rangle\langle m_0\Gamma_0 y\|Q^s\|m\Gamma x\rangle, \quad (1)$$

where $(m - m_0) \equiv \Delta k$ is the net difference between the number of single-particle creation and annihilation operators in P^r.

By multiplying both sides of the equation with the appropiate $6j$-symbol, phase factor, and statistical weight, we obtain

$$\sum_t(-1)^{2\Gamma+t}[\Gamma_0'][t]^{\frac{1}{2}}\begin{Bmatrix}\Gamma & r & \Gamma_0' \\ s & \Gamma & t\end{Bmatrix}\langle m\Gamma x\|(P^r \times Q^s)^t\|m\Gamma x\rangle$$

$$= \sum_y \langle m\Gamma x\|P^r\|m_0\Gamma_0' y\rangle\langle m_0\Gamma_0' y\|Q^s\|m\Gamma x\rangle, \quad (2)$$

after summing over t on the right hand side and making use of the orthogonality relation of $6j$-symbols given by eq. (B-15a). Since the reduced matrix elements of P^r and Q^s are numbers, we can interchange their positions and obtain the relation

$$\sum_x \langle m_0\Gamma_0 y\|Q^s\|m\Gamma x\rangle\langle m\Gamma x\|P^r\|m_0\Gamma_0 y\rangle = \sum_{t'}(-1)^{2\Gamma_0+t'}[\Gamma][t']^{\frac{1}{2}}$$

$$\times \begin{Bmatrix}\Gamma_0 & r & \Gamma \\ s & \Gamma_0 & t'\end{Bmatrix}\langle m_0\Gamma_0 y\|(Q^s \times P^r)^{t'}\|m_0\Gamma_0 y\rangle, \quad (3)$$

in analogy with eq. (2) above.

Alternatively we can sum over x in eq. (1) and express the reduced trace, i.e., sum of the reduced matrix elements, of $(P^r \times Q^s)^t$ in the form

$$\ll m\Gamma\|(P^r \times Q^s)^t\|m\Gamma \gg = \sum_x \langle m\Gamma x\|(P^r \times Q^s)^t\|m\Gamma x\rangle$$

$$= \sum_{\Gamma_0 xy}(-1)^{2\Gamma+t}[t]^{\frac{1}{2}}\begin{Bmatrix}\Gamma & r & \Gamma_0 \\ s & \Gamma & t\end{Bmatrix}\langle m_0\Gamma_0 y\|Q^s\|m\Gamma x\rangle\langle m\Gamma x\|P^r\|m_0\Gamma_0 y\rangle$$

$$= \sum_{\Gamma_0 t'y}(-1)^{t-t'+2\Gamma-2\Gamma_0}[\Gamma][tt']^{\frac{1}{2}}\begin{Bmatrix}\Gamma & r & \Gamma_0 \\ s & \Gamma & t\end{Bmatrix}\begin{Bmatrix}\Gamma_0 & r & \Gamma \\ s & \Gamma_0 & t'\end{Bmatrix}$$

$$\times \langle m_0\Gamma_0 y\|(Q^s \times P^r)^{t'}\|m_0\Gamma_0 y\rangle$$

$$= \sum_{\Gamma_0 t'}(-1)^{t-t'+2\Gamma-2\Gamma_0}[\Gamma][tt']^{\frac{1}{2}}\begin{Bmatrix}\Gamma & r & \Gamma_0 \\ s & \Gamma & t\end{Bmatrix}\begin{Bmatrix}\Gamma_0 & r & \Gamma \\ s & \Gamma_0 & t'\end{Bmatrix}$$

$$\times \ll m_0\Gamma_0\|(Q^s \times P^r)^{t'}\|m_0\Gamma_0 \gg, \quad (4)$$

where the second line is arrived at by using eq. (3) and the final result by converting the matrix elements into a trace over the label y. The appearance of the equation can be simplified by defining a single symbol

$$T(rs; tt'; \Gamma\Gamma_0) \equiv (-1)^{t-t'+2\Gamma-2\Gamma_0}[\Gamma][tt']^{\frac{1}{2}} \begin{Bmatrix} \Gamma & r & \Gamma_0 \\ s & \Gamma & t \end{Bmatrix} \begin{Bmatrix} \Gamma_0 & r & \Gamma \\ s & \Gamma_0 & t' \end{Bmatrix}, \quad (5)$$

for all the angular momentum factors.

We can change the order of Q^s and P^r on the right-hand-side of eq. (4) by making use of either the commutation or anticommutation relation between Q^s and P^r depending on whether we have an even or odd number of single-particle creation and annihilation operators in each. In angular momentum coupled form, we can write

$$[Q^s, P^r]^t_{\pm} \equiv (Q^s \times P^r)^t \pm (-1)^{r+s-t}(P^r \times Q^s)^t. \quad (6)$$

If the particle rank of the operator $\hat{O}^t = (P^r \times Q^s)^t$ is k, the commutator $[P^r, Q^s]^t_{\pm}$ has particle rank $(k-1)$, and is therefore a simpler operator than $(P^r \times Q^s)^t$ itself.

We now obtain the propagation equation by incorporating eqs. (5) and (6) into eq. (4):

$$\ll m\Gamma \| (P^r \times Q^s)^t \| m\Gamma \gg$$
$$= \sum_{\Gamma_0 t'} T(rs; tt'; \Gamma\Gamma_0) \{ \ll m_0\Gamma_0 \| [Q^s, P^r]^{t'}_{\pm} \| m_0\Gamma_0 \gg$$
$$\mp (-1)^{r+s-t} \ll m_0\Gamma_0 \| (P^r \times Q^s)^{t'} \| m_0\Gamma_0 \gg \}. \quad (7)$$

It relates the trace of an operator $\hat{O}^t$ in m-particle space to its traces in m_0-particle space. The propagation process therefore depends on how the operator $\hat{O}^t$ is partitioned into P^r and Q^s. If $\Delta k = 0$, we are relating the traces within the space of the same particle number m.

Alternatively, we can partition $\hat{O}^t$ into P^r and Q^s with $\Delta k \neq 0$. For example, if

$$\hat{O}^t = (A^\rho \times B^\rho)^t,$$

we can have

$$P^r = A^\rho, \qquad Q^s = B^\rho.$$

Hence $\Delta k = 1$, and

$$[B^\rho, A^\rho]^t_{+} = [\rho]^{\frac{1}{2}}\delta_{t0}.$$

Using eq. (7), the fixed-JT trace of $(A \times B)^t$ is propagated from the $(m-1)$-particle space to the m-particle space by

$$\ll m\Gamma \| (A^\rho \times B^\rho)^t \| m\Gamma \gg = \sum_{\Gamma_0 t'} T(\rho\rho; tt'; \Gamma\Gamma_0) \{ [\rho\Gamma_0]^{\frac{1}{2}} d_{(m-1)\Gamma_0}$$
$$- (-1)^{2\rho-t} \ll (m-1)\Gamma_0 \| (A^\rho \times B^\rho)^{t'} \| (m-1)\Gamma_0 \gg \}, \quad (8)$$

where we have made use of the relation between the trace of a unity operator and $d_{m\Gamma}$,

$$\ll m\Gamma\|1\|m\Gamma\gg= [\Gamma]^{\frac{1}{2}}d_{m\Gamma}.$$

We can start the process from $m = 1$ by calculating explicitly the trace of $(A \times B)^t$ in a j-orbit. This can be done, for example, using eq. (C-20), and the relations given by eqs. (C-15) and (C-16) between fractional parentage coefficient and the matrix elements of A^ρ and B^ρ.

For operators with higher particle ranks we can partition $\hat{O}^t$ in different ways. For simplicity, we shall discuss the choice of $\Delta k = 1$ only. For the operator $(((A \times A)^{\Delta_1} \times B)^{\Delta_2} \times B)^t$ we shall take

$$P^r = ((A \times A)^{\Delta_1} \times B)^{\Delta_2}, \quad \text{and} \quad Q^s = B^\rho.$$

The anticommutator is

$$\left[B^\rho, ((A \times A)^{\Delta_1} \times B)^{\Delta_2}\right]^t_+ = (-1)^{2\rho+t}2\,[\Delta_1\Delta_2]^{\frac{1}{2}} \left\{\begin{matrix} \rho & \Delta_1 & \rho \\ \rho & t & \Delta_2 \end{matrix}\right\}(A \times B)^t.$$

$$(9)$$

Fixed-JT traces of the two-body elementary operator can therefore be propagated starting from $m = 2$ and making use of the traces of the one-body operator calculated using eq. (8).

The three-body operator can be recoupled into the form

$$\left(A \times \left((A \times A)^{\Delta_1} \times (B \times (B \times B)^{\Delta_2})^{\Delta_3}\right)^{\Delta_4}\right)^t.$$

By taking

$$P^r = A^\rho \quad \text{and} \quad Q^s = \left((A \times A)^{\Delta_1} \times (B \times (B \times B)^{\Delta_2}x)^{\Delta_3}\right)^{\Delta_4},$$

the anticommutator between them contains two two-body operators,

$$\left[\left((A \times A)^{\Delta_1} \times (B \times (B \times B)^{\Delta_2})^{\Delta_3}\right)^{\Delta_4}, A^\rho\right]^t_+ = -\sum_\gamma (-1)^{\Delta_1+\Delta_3-\rho+t}$$

$$[\Delta_3\Delta_4]^{\frac{1}{2}}\left\{\begin{matrix} \Delta_1 & \Delta_3 & \Delta_4 \\ \rho & t & \gamma \end{matrix}\right\}\left\{\delta_{\gamma\Delta_2}((A \times A)^{\Delta_1} \times (B \times B)^{\Delta_2})^t\right.$$

$$\left. + (-1)^\gamma 2[\Delta_2\gamma]^{\frac{1}{2}}\left\{\begin{matrix} \rho & \rho & \gamma \\ \rho & \Delta_3 & \Delta_2 \end{matrix}\right\}((A \times A)^{\Delta_1} \times (B \times B)^\gamma)^t\right\}. \quad (10)$$

The fixed-JT traces of the three-body operator can therefore be propagated starting from $m = 3$ and using the traces of the two-body operator from eq. (9).

For the scalar four-body operator, we shall take

$$P^r = \left(\left(\left((A \times A)^{\Delta_1} \times (B \times B)^{\Delta_1}\right)^0 \times \left((A \times A)^{\Delta_2} \times B\right)^\rho\right)^\rho\right)^\rho,$$

and $Q^s = B^\rho$. The anticommutator contains one-, two- as well as three-body operators,

$$\left[B^\rho, \left(\left(\left((A \times A)^{\Delta_1} \times (B \times B)^{\Delta_1}\right)^0 \times \left((A \times A)^{\Delta_2} \times B\right)^\rho\right)^\rho\right)\right]_+^t$$

$$= -4\delta_{\Delta_1 \Delta_2}[\rho]^{\frac{1}{2}}(-1)^t \begin{Bmatrix} \rho & \rho & \Delta_1 \\ \rho & \rho & t \end{Bmatrix} (A \times B)^t$$

$$- 4[\Delta_2\rho]^{\frac{1}{2}} \begin{Bmatrix} \rho & \rho & \Delta_2 \\ \rho & \rho & t \end{Bmatrix} \sum_\gamma [\gamma]^{\frac{1}{2}} \begin{Bmatrix} \Delta_1 & \rho & \rho \\ \rho & t & \gamma \end{Bmatrix} \left((A \times A)^{\Delta_1} \times (B \times B)^\gamma\right)^t$$

$$- 8[\Delta_1 \Delta_2 \rho]^{\frac{1}{2}} \sum_{\gamma_1 \gamma_2 \eta} [\eta][\gamma_1 \gamma_2]^{\frac{1}{2}} \begin{Bmatrix} \rho & \rho & \Delta_1 \\ \rho & \eta & t \end{Bmatrix} \begin{Bmatrix} \rho & \rho & \gamma_2 \\ \gamma_1 & t & \eta \end{Bmatrix} \begin{Bmatrix} \Delta_1 & \rho & \rho \\ \eta & \gamma_1 & \rho \\ \rho & \rho & \Delta_2 \end{Bmatrix}$$
$$\times \left((A \times A)^{\gamma_1} \times (B \times B)^{\gamma_2}\right)^t$$

$$- 2\left[\frac{\Delta_2 \rho}{\Delta_1}\right]^{\frac{1}{2}} \begin{Bmatrix} \rho & \rho & \Delta_2 \\ \rho & \rho & t \end{Bmatrix} \sum_{\gamma_1 \gamma_2} (-1)^{\rho - \gamma_2} [\gamma_1 \gamma_2]^{\frac{1}{2}} \begin{Bmatrix} \rho & \rho & t \\ \gamma_1 & \gamma_2 & \Delta_1 \end{Bmatrix}$$
$$\times \left((A \times (A \times A)^{\Delta_1})^{\gamma_1} \times (B \times (B \times B)^{\Delta_2})^{\gamma_2}\right)^t$$

$$- 2[\rho]^{\frac{1}{2}} \sum_{\gamma_1 \gamma_2} [\gamma_1 \gamma_2]^{\frac{1}{2}} (-1)^{\rho - \gamma_2} \begin{Bmatrix} \rho & \Delta_1 & \rho \\ \Delta_2 & \rho & \rho \\ \gamma_1 & \gamma_2 & t \end{Bmatrix}$$
$$\times \left((A \times (A \times A)^{\Delta_2})^{\gamma_1} \times (B \times (B \times B)^{\Delta_1})^{\gamma_2}\right)^t. \qquad (11)$$

Since the four-body elementary operator is not in normal order form, it contains $k = 2$ and $k = 3$ parts. The propagation must therefore be started from $m = 2$.

The method of step-by-step propagation is not restricted to traces in spin-isospin averaging. In fact, if the recoupling coefficients are available, the iterative propagation method can be applied to other group averages as well.

References

Ayik, S., and Ginocchio, 1974, *Nucl. Phys.* **A221**, 285

Barrett, B.R., and Kirson, M.W., 1970, *Nucl. Phys.* **A148**, 145

Bethe, H.A., 1937, *Rev. Mod. Phys.* **9**, 69

Bethe, H.A., Brown, G.E., Applegate, J., and Lattimer, J.M., 1979, *Nucl. Phys.* **A324**, 487

Betrand, F., 1981, *Nucl. Phys.* **A354**, 129c

Biedenharn, L.C., 1962, *Phys. Lett.* **3**, 69

Bilpuch, E.G., Lane, A.M.. Mitchell, G.E. and Moses, J.D., 1976, *Phys. Rep. C* **28**, 145

Bohigas, O., and Giannoni, M.J., 1975, *Ann. Phys.* **89**, 393

Bohr, A., and Mottelson, B.R., 1969, *Nuclear Structure*, vol. I (Benjamin, New York)

Brink, D.M., and Satchler, G.R., 1968, *Angular Momentum* (Clarendon Press, Oxford)

Brody, T.A., 1973, *Lett. Nuovo Cimento* **7**, 482

Brody, T.A., Cota, E., Flores, J., and Mello, P.A., 1976, *Nucl. Phys.* **A259**, 87

Brody, T.A., Flores, J., French, J.B., Mello, P.A., Pandey, A., and Wong, S.S.M., 1981, *Rev. Mod. Phys.* **53**, 385

Chang, B.D., Vincent, C.M., and Wong, S.S.M., 1984, *Phys. Rev.* **C30**, 1055

Chang, B.D., and Wong, S.S.M., 1978, *Nucl. Phys.* **A294**, 19

Chang, B.D., and Wong, S.S.M., 1979, *Comput. Phys. Commun.* **18**, 35

Chang, B.D., and Wong, S.S.M., 1980, *Comput. Phys. Commun.* **20**, 191

Chang, B.D., Draayer, J.P., and Wong, S.S.M., 1982, *Comput. Phys. Commun.* **28**, 41

Chang, F.S., French, J.B., and Thio, T.H., 1971, *Ann. Phys.* **66**, 137

Chung, W., 1976, *thesis,* Michigan State University

Coceva, C., and Stefanon, M., 1979, *Nucl. Phys.* **A315**, 1

Countee, C.R., Draayer, J.P., Halemane, T.R., and Kar, K., 1981, *Nucl. Phys.* **A356**, 1

Cramer, H., 1946, *Mathematical Methods of Statistics* (Princeton Univ. Press, Princeton)

De-Shalit, A., and Talmi, I., 1963, *Nuclear Shell Theory* (Academic Press, New York)

Dilg, W., Schantl, W., Vonach, H. and Uhl, M., 1973, *Nucl. Phys.* **A217**, 269

Draayer, J.P., French, J.B., Prasad, M., Potbhare, V., and Wong, S.S.M., 1975, *Phys. Lett.* **57B**, 130

Draayer, J.P., French, J.B., and Wong, S.S.M., 1977, *Ann. Phys.* **106**, 472; 503

Draayer, J.P., and Rosensteel, G., 1983, *Phys. Lett.* **124B**, 281

Dyson, F.J., and Mehta, M.L., 1963, *J. Math. Phys.* **4**, 701

Eisenberg, J.M., and Greiner, W., 1972, *Microscopic Theory of the Nucleus* (North-Holland, Amsterdam)

Elliot, J.P., 1958, *Proc. Roy. Soc.* **A245**, 128; 562

Elliot, J.P., 1963, in *Selected Topics in Nuclear Theory*, edited by F. Janouch (Intern. At. Energy Agency, Vienna)

Feller, W., 1971, *An Introduction to Probability Theory and Its Applications*, vol. II (Wiley, New York)

Flowers, B.H., 1952, *Proc. Roy. Soc.* **A212**, 248

French, J.B., 1969, in *Isospin in Nuclear Physics*, edited by D.H. Wilkinson (North-Holland, Amsterdam)

French, J.B., and Draayer, J.P., 1979, in *Group Theoretical Methods in Physics*, edited by W. Beiglbock, A. Bohm and E. Takasugi (Springer, Berlin)

French, J.B., and Kota, V.K.B., 1983, *Phys. Rev. Lett.* **51**, 2183

French, J.B., and Wong, S.S.M., 1970, *Phys. Lett.* **33B**, 449

Garrison, J.D., 1964, *Ann. Phys.* **30**, 269

Gervois, A., 1972, *Nucl. Phys.* **A184**, 507

Goode, P., and Koltun, D.S., 1975, *Nucl. Phys.* **A243**, 44

Grimes, S.M., Poppe, C.H., Wong, C., and Dalton, B.J., 1978, *Phys. Rev.* **C18**, 1100

Halemane, T.R., Kar, K., and Draayer, J.P., 1978, *Nucl. Phys.* **A311**, 301

Halemane, T.R., 1981, *Z. Phys.* **A301**, 177

Hamermesh, M., 1962, *Group Theory and Its Application to Physical Problems* (Addison-Wesley, Reading, Mass.)

Haq, R.U., and Wong, S.S.M., 1979, *Nucl. Phys.* **A327**, 314

Haq, R.U., and Wong, S.S.M., 1980, *Phys. Lett.* **93B**, 357

Haq, R.U., and Wong, S.S.M., 1982, *Can. J. Phys.* **60**, 1502

Haq, R.U., Pandey, A., and Bohigas, O., 1982, *Phys. Rev. Lett.* **48**, 1086

Hillman, M., and Grover, J.R., 1969, *Phys. Rev.* **185**, 1303

Hugenholtz, N.M., 1957, *Physica* **23**, 481

Huizenga, J.R., and Moretto, L.G., 1972, *Ann. Rev. Nucl. Sci.* **22**, 427

Jacquemin, C., 1980, in *Theory and Applications of Moment Methods in Many-Fermion Systems*, edited by B.J. Dalton, S.M. Grimes, J.P. Vary and S.A. Williams (Plenum, New York)

Jacquemin, C., and Spitz, S., 1979, *Z. Phys.* **A290**, 251

Kahn, P.B., and Rosenzweig, N., 1969, *Phys. Rev.* **187**, 1193 and earlier works cited therein

Kuo, T.T.S., 1967, *Nucl. Phys.* **A103**, 71

Liou, H.I., Camarda, H.S., Wynchank, S., Slagowitz, M., Hacken, G., Rahn, F., and Rainwater, J., 1972, *Phys. Rev.* **C5**, 974

Lougheed, G.D., and Wong, S.S.M., 1975, *Nucl. Phys.* **A243**, 215

Lynn, J.E., 1968, *The Theory of Neutron-Resonance Reactions* (Clarendon, Oxford)

Mehta, M.L., and Gaudin, M., 1960, *Nucl. Phys.* **18**, 420

Mon, K.K., and French, J.B., 1975, *Ann. Phys.* **95**, 90

Monahan, J.E., and Rosenzweig, N., 1972, *Phys. Rev.* **C5**, 1078

Mugambi, P.E., 1970, *thesis*, University of Rochester

Ng, W.Y., and Wong, S.S.M., 1976, *Can. J. Phys.* **54**, 2367

Pandey, A., 1979, *Ann. Phys.* **119**, 170

Parikh, J.C., 1972, *Phys. Lett.* **41B**, 468

Parikh, J.C., 1973, *Ann. Phys.* **76**, 202

Parikh, J.C., 1978, *Group Symmetries in Nuclear Structure* (Plenum, New York)

Potbhare, V., 1977, *Nucl. Phys.* **A289**, 373

Preedom, B.M., and Wildenthal, B.H., 1972, *Phys. Rev.* **C6**, 1633

Quesne, C., 1975, *J. Math. Phys.* **16**, 2427

Quesne, C., 1976, *J. Math. Phys.* **17**, 1452

Quesne, C., and Spitz, S., 1974, *Ann. Phys.* **85**, 115

Riordan, J., 1968, *Combinatorial Identities* (Wiley, New York)

Verbaarschot, J.J.M., and Brussaard, P.J., 1981, *Phys. Lett.* **102B**, 201

von Neumann, J., and Wigner, E.P., 1929, *Z. Phys.* **30**, 467

Wigner, E.P., 1937, *Phys. Rev.* **51**, 106

Wigner, E.P., 1955, *Ann. Math.* **62**, 548

Wigner, E.P., 1967, *SIAM Review* **9**, 1

Wigner, E.P., 1972, in *Statistical Properties of Nuclei*, edited by J.B. Garg (Plenum Press, New York)

Wildenthal, B.H., and Chung, W., 1979, in *Mesons in Nuclei*, edited by M. Rho and D.H. Wilkinson (North-Holland, Amersterdam)

Wong, S.S.M., 1978, *Nucl. Phys.* **A295**, 275

Wong, S.S.M., and French, J.B., 1972, *Nucl. Phys.* **A198**, 188

Wong, S.S.M., and Lougheed, G.D., 1978, *Nucl. Phys.* **A295**, 289

Wong, S.S.M., Zhao, E., and Zhu, X., 1980, *Astrop. Journ. (Letters)* **242**, L173

Notation

A, nucleon number

Angular momentum recoupling coefficients,

$\langle rpsq|tm \rangle$, Clebsch-Gordan coefficient

$\begin{pmatrix} j_1 & j_2 & j_3 \\ m_1 & m_2 & m_3 \end{pmatrix}$, 3$j$-symbol

$W(j_1 j_2 J j_3; J_{12} J_{23})$, Racah coefficient

$\begin{Bmatrix} j_1 & j_2 & J_{12} \\ j_3 & J & J_{23} \end{Bmatrix}$, 6$j$-symbol

$\begin{Bmatrix} j_1 & j_2 & J_{12} \\ j_3 & j_4 & J_{34} \\ J_{13} & J_{24} & J \end{Bmatrix}$, 9$j$-symbol

$a_r^\dagger$, single-particle creation operator for state r

a_r, single-particle annihilation operator for state r

A^r, spherical tensor single-particle creation operator for orbit r

B^r, spherical tensor single-particle annihilation operator for orbit r

C, centroid

D, average level spacing

d, dimension of a subspace

$D_L^p(\alpha)$, the αth way to contract $\hat{O}$ with p left-contractions

$D_R^p(\alpha)\hat{O}$, the αth way to contract $\hat{O}$ with p right-contractions

$\mathcal{D}_{nm}^l(\alpha\beta\gamma)$, rotation matrix

$\overline{E}$, average trace of H

$G_p(E) = \langle E|\hat{O}^\dagger H^p \hat{O}|E \rangle$, energy-weighted sum rule of order p

H, Hamiltonian

  ϵ_r, single-particle energy of orbit r

  ϵ_{rs}, off-diagonal single-particle energy

  $\tilde{\epsilon}_r$, traceless single-particle energy

  $H(1)$, one-body Hamiltonian

  $H(2)$, two-body Hamiltonian

  $H(0,0)$, unitary rank zero part of H

  $H(1,0)$, unitary rank one part of H

  $H(2,0)$, unitary rank two part of H

  $W_{rstu}^\Gamma = \langle rs\Gamma|V|tu\Gamma \rangle$, normalized and antisymmetrized two-body matrix element

  $\overline{W}$, average two-body matrix element

  $\overline{W}_{rs}$, induced single-particle energy (scalar)

  $\mathcal{W}_{rstu}^\Gamma$, two-body matrix element of the unitary rank-two part of H (scalar)

  $\overline{W}(r,s)$, induced zero-body energy (scalar)

  $\overline{W}(r,s;t)$, induced single-particle energy (configuration)

  V_{rstu}^Γ, two-body matrix element of the unitary rank-two part of H (configuration)

  V, two-body interaction

$I(x)$, density of states, normalized to d

   $\rho(x)$, density of states normalized to unity

J, spin

k, particle rank

$K(E) = \langle E|\hat{K}|E\rangle$, expectation value of $\hat{K}$ at E

L, angular momentum

   M, projection of L on the quantization axis

M_μ, moment of order μ

   $\mathcal{M}_\mu$, central moment of order μ

$\hat{n}$, number operator

   $\hat{n}_r$, number operator for orbit r

N, total number of single-particle states in the space

$\binom{n}{m}$, binomial coefficient

$\hat{O}_{\text{diag}}$, diagonal part of $\hat{O}$

$\hat{O}_{\text{tr}}$, trace-equivalence operator for $\hat{O}$

$\ll \hat{O} \gg = \sum_i \langle i|\hat{O}|i\rangle$, trace of $\hat{O}$

$\langle \hat{O} \rangle = d^{-1} \ll \hat{O} \gg$, average trace of $\hat{O}$

$\overline{O}$, adjoint of operator $\hat{O}$

$\vdots\hat{O}\vdots$, normal-ordered product of $\hat{O}$ with respect to particle vacuum

$\circ\hat{O}\circ$, normal-ordered product of $\hat{O}$ with respect to hole vacuum

O_t^p, fully contracted quantity for $\hat{O}$ with t left-contractions and p contractions in total

$P_\mu(x)$, polynomial of order μ

$\overline{Q}$, ensemble average value of Q

$R(E', E)$, excitation strength from state $|E\rangle$ to state $|E'\rangle$

$S = E_{i+1} - E_i$, nearest neighbour spacing

   $S_i^k = E_{i+k+1-E_i}$, level spacing of order k

T, isospin

T_q^p, U_q^p, spherical tensors of rank p and projection q

U_{rs}^Δ, Racah unit tensor

v_{rs}^λ, multipolarity λ one-body operator single-particle value between orbits r and s

$Y_m^l(\theta', \phi')$, spherical harmonic

$Z^\Gamma(mx)$, state creation operator

   $\overline{Z^\Gamma}(mx)$, adjoint of $Z^\Gamma(mx)$

$[\Gamma] = (2J + 1)(2T + 1)$, statistical weight of spin-isospinfactor

(μ, ν), unitary rank

$\rho(x)$, density of states normalized to unity

σ^2, variance

σ, width

$\varsigma_{rs} = \dfrac{1}{\sqrt{1+\delta_{rs}}}$, antisymmetrization factor

Index